Graphing Calculator

for NYS Math B ...

and

Beyond

by

Kathleen Noftsier

Note for Librarians: A cataloguing record for this book is available from Library and Archives
Canada at www.collectionscanada.ca/amicus/index-e.html
ISBN 1-4251-0216-6

*Trafford's print shop runs on "green energy" from solar, wind and other environmentally-friendly
power sources.*

Offices in Canada, USA, Ireland and UK

Book sales for North America and international:
Trafford Publishing, 6E–2333 Government St.,
Victoria, BC V8T 4P4 CANADA
phone 250 383 6864 (toll-free 1 888 232 4444)
fax 250 383 6804; email to orders@trafford.com
Book sales in Europe:
Trafford Publishing (UK) Limited, 9 Park End Street, 2nd Floor
Oxford, UK OX1 1HH UNITED KINGDOM
phone +44 (0)1865 722 113 (local rate 0845 230 9601)
facsimile +44 (0)1865 722 868; info.uk@trafford.com
Order online at:
trafford.com/06-1973

10 9 8 7 6 5 4 3 2 1

Table of Contents

Graphing Calculator for Math B and Beyond

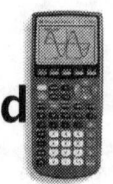

This workbook is designed to be completed with a TI-83+/TI-84+ graphing calculator. If you do not have one, you might be able to borrow one from the school. It is advisable to leave this calculator at home if your teacher has a classroom set available, as **you will be responsible for the replacement cost if you are unable to return it on the last day of the course**.

This book was written for students that have completed the basic graphing calculator course and assumes that the student is already familiar with the calculator. If this is not the case, some lessons may need some extra explanation from the instructor or you can check the calculator's manual for additional directions.

The TI-83+/TI-84+ has two identification numbers.
1. The first is on the outside. Find the number on the back of the calculator you will be responsible for (yours or the school's) and record it here:

    _____

2. The second is an internal number. To find it:
    a. Turn the calculator on.

    b. Go to Memory (**2ⁿᵈ, +**).

    c. Choose **1:About**

    d. Look for **ID:**

    e. Write the calculator's ID# here: _____

    f. Your signature: _____

You should also note the number under the product name. This is the number of the current operating system. The operating system should be updated so that the most current system is loaded to ensure the best performance. Also, as new applications are developed, the programming necessary to make them run is included in the updated operating system. If you don't have the latest operating system you might not be able to load these new applications to your calculator.

At this time, the most current OS is: _____

If your calculator has an OS lower than this it should be updated at the earliest opportunity.
If you have a TI Graph Link you may download programs from the TI website or a computer that has the program or application, or with a unit-to-unit link from another calculator. Do not try to install non-TI programs to a TI calculator. You can damage the operating system of the calculator. In most cases it can be reinstalled, but you will lose all information on the calculator. **Don't take the chance!!**

Regents Connection B

Included in this workbook are   pages.
These are collections of Math B regents questions related to the topic immediately preceding them. They may or may not be appropriate for use on the graphing calculator. Part of your training in this course is to determine when calculator use is appropriate, so use your best judgment. They may be omitted at your teacher's discretion if you are taking this course only for math credit and are not anticipating taking the Math B regents exam (although they still make great practice for the skills learned in this book!)

Also, in order to complete the entire course in 20-weeks, it will be necessary for you to complete some of these outside of class. Time has **_not_** been allowed as in the preceding course, *Graphing Calculator for Math A and Beyond*, to complete all work in class. Because Math B students are required to have access to graphing calculators, it is assumed that you have one to use outside the classroom and homework will be assigned as part of this course.

Good Luck and Happy Calculating!

This lesson will review how to handle basic operations on the TI-83+/TI-84+.

That may sound too simple, but many students lose points on the Math B regents or other exams because they are not careful with the basics.

To illustrate the importance of computing carefully we will use the quadratic formula. Every math B student should have this formula memorized. The next hurdle is to enter the parts into the calculator correctly. MANY STUDENTS DON'T!!

The formula:

$$x = \frac{-b \pm \sqrt{b^2 - 4ac}}{2a}$$

The problem with this formula is that there are so many places to make a mistake if you are not familiar with exactly how to enter it in the calculator.

Let's look at the "Big Four". If you have the wrong answer you probably made one of the following mistakes:

1. The most frequently made mistake is made when "b" is negative. Find -7^2 on the calculator. Maybe you don't make this mistake, but many students say this is -49. Why? The calculator reads -7^2 as $-1 \cdot 7^2$. What does the order of operations say the answer to this expression is? _____ You really want $(-7)^2$. (I tell my students to square "b" in their heads; they're more likely to get the problem right!)

2. The next mistake is related to the first, misuse of the negative symbol and the subtraction key. You should be subtracting between $-b$ and $\sqrt{b^2 - 4ac}$, and between b^2 and $4ac$ under the radical. All others should use the negative symbol.

3. Mistake number three occurs if a student is entering the entire expression and forgets that the calculator can't "see" where the square root ends. Don't forget to use parentheses after the "c" to let the calculator know that this is the last term under the square root.

4. The last mistake is a grouping error. Again, it only occurs when entering the entire expression at once. The calculator can't determine what part of the expression is in the numerator unless you tell it, and the denominator must divide the entire numerator.

Mistakes (3) and (4) won't always apply; if an answer in simplest radical form is required the decimal answer found by entering the entire formula will not be appropriate. But these will still make good practice.

Quadratic Formula, Etc.

<u>June '02, #19:</u> The roots of the equation $2x^2 - x = 4$ are

     (1) real and irrational
     (2) real, rational, and equal
     (3) real, rational, and unequal
     (4) imaginary

<u>Aug '01, #6:</u> The roots of the equation $x^2-3x-2=0$ are

     (1) real, rational, and equal
     (2) real, rational, and unequal
     (3) real, irrational, and unequal
     (4) imaginary

<u>Jan '02, #1:</u> The roots of a quadratic equation are real, rational, and equal when the discriminant is

     (1) –2   (2) 2   (3) 0   (4) 4

<u>June '01, #3:</u> Jacob is solving a quadratic equation. He executes a program on his graphing calculator and sees that the roots are real, rational, and unequal. This information indicates to Jacob that the discriminant is

(1) zero   (2) negative   (3) a perfect square   (4) not a perfect square

<u>June '03, #7:</u> The roots of the equation $ax^2 + 4x = -2$ are real, rational, and equal when a has a value of

     (1) 1   (2) 2   (3) 3   (4) 4

<u>Aug '02, #11:</u> Which equation has imaginary roots?

(1) $x^2-1=0$ (3) $x^2+x+1=0$
(2) $x^2-2=0$ (4) $x^2-x-1=0$

<u>Aug '02, #17:</u> If the sum of the roots of $x^2+3x-5=0$ is added to the product of its roots, the result is

(1) 15   (2) –15   (3) –2   (4) –8

<u>Jan '03, #13:</u> If the roots of $ax^2+bx+c=0$ are real, rational, and equal, what is true about the graph of the function $y=ax^2+bx+c$?

(1) It intersects the x-axis in two distinct points.
(2) It lies entirely below the x-axis.
(3) It lies entirely above the x-axis.
(4) It is tangent to the x-axis.

<u>Aug '03, #20:</u> In the equation $ax^2+6x-9=0$, imaginary roots will be generated if

(1) $-1<a<1$ (2) $a<1$, only   (3) $a>-1$, only   (4) $a<-1$

<u>Jan '04, #16:</u> Which statement must be true if a parabola represented by the equation $y=ax^2+bx+c$ does not intersect the x-axis?

(1) $b^2-4ac=0$
(2) $b^2-4ac<0$
(3) $b^2-4ac>0$, and b^2-4ac is a perfect square.
(4) $b^2-4ac>0$, and b^2-4ac is not a perfect square.

Properties of Real Numbers

June '02, #7: Which statement is true for all real number values of x?

(1) $|x-1| > 0$ (2) $|x-1| > (x-1)$ (3) $\sqrt{x^2} = x$ (4) $\sqrt{x^2} = |x|$

Sample #9: Which is the correct arrangement of these terms in order of value, from smallest to greatest:

(1) $3\sqrt{2}$, $4\frac{1}{8}$, $|-4.24|$, $\sqrt[3]{75}$

(2) $\sqrt[3]{75}$, $|-4.24|$, $4\frac{1}{8}$, $3\sqrt{2}$

(3) $4\frac{1}{8}$, $\sqrt[3]{75}$, $|-4.24|$, $3\sqrt{2}$

(4) $4\frac{1}{8}$, $|-4.24|$, $\sqrt[3]{75}$, $3\sqrt{2}$

Sample #21: Show that the following can be ordered from smallest to largest for all $x > 1$. Describe the method you used and state the correct order.

$$1, \ x, \ \sqrt{x}, \ \frac{1}{x}, \ \frac{1}{\sqrt{x}}$$

Jan '02, #17: The value of $\left(\dfrac{3^0}{27^{\frac{2}{3}}}\right)^{-1}$ is

(1) -9 (2) 9 (3) $-\dfrac{1}{9}$ (4) $\dfrac{1}{9}$

Aug '03, #26: Tom scored 23 points in a basketball game. He attempted 15 field goals and 6 free throws. If each successful field goal is 2 points and each successful free throw is 1 point, is it possible he successfully made all 6 of his free throws? Justify your answer.

 # Equation "Checker"

Previously known as Equation Solver, it is more appropriate for Math B to call it Equation Checker.

Equations of more complex functions require more care in order to solve them correctly both on paper and in Equation Solver. Knowing what a reasonable solution should look like and knowing the characteristics of a particular function are essential.

Through this lesson, it will be assumed that you do understand the traditional methods of solving these equations. It is your responsibility to be sure that you can solve each equation using paper and pencil methods.  **_Relying on Equation Solver will not guarantee success on the Math B regents or in further math courses_**.

Equation "Checker" is found by pressing $\boxed{\text{MATH}}$ and choosing **0:Solver**.

Keys to success:

 1. Only equations with one unknown can be solved.
    a. No "=" sign - not an equation!
    b. No variable - no unknown to solve for!

2. Every equation must be set equal to zero. The quick way to do this for calculator purposes is to change the equal sign to subtraction and group the side to the right of the equal sign with parentheses. (Note: the parentheses are not necessary if the right side is a monomial.)

3. Be sure you are in the correct MODE for trigonometric equations.

4. For cyclic functions, choose an initial value for x that is in the cycle you would expect to find the solution or at least in the interval stated in the question if one is given. (This may not make sense until you have studied the graphs of trigonometric functions.)

5. Know how many solutions are possible.

6. If you change the form of the original equation to make it work better in solver, check the answer in the original to be sure the equation is defined at this point. (i.e. If you cross-multiply a proportion to make it work without giving a divide by zero or no sign change error, check the original ratios to determine if they are defined at the value the calculator tells you is the solution.)

7. If a solution seems unreasonable, try changing the value of x and solving again to see if you get the same result. Examples of false "solutions" will be looked at when we study logarithmic and exponential functions.

8. In addition to using parentheses on the right side of the equation, be sure to group
    a. numerators of fractions
    b. denominators of fractions
    c. exponents that are sums, differences, products, or quotients
    d. degrees (or radians)
    e. parts of an expression under a radical

9. Round repeating zeros or repeating nines.

10. Remember, if you need to use the variable in an expression, you can use the variable itself as that value on the home screen.

11. Having trouble clearing the last equation? Be sure you have not cleared the variable. If you have, the calculator will not let you move up to the equation. Just pick any value for the variable. The calculator doesn't care what the value of x (or any other variable) is, but it will not allow you to leave it "empty".

Practice:

Show the algebraic solution for each equation then check with Solver.

1. $\sqrt{2x^2 - 3} = x + 2$

2. $\sqrt{x+6} + 2x = 3 + x$

3. $4 + x = 3 - \sqrt{x+2}$

4. $\dfrac{x+4}{3} = \dfrac{4}{x}$

5. $\dfrac{5}{2x} = \dfrac{x-3}{4}$

6. $0 = -x^2 - 5x + 12$

7. $0 = 3x^2 + 2x - 5$

8. $0 = 0.05x^2 - 0.3x - 0.09$

June '02, #5: The path of a rocket is represented by the equation $y = \sqrt{25 - x^2}$. The path of a missile designed to intersect the path of the rocket is represented by the equation $x = \frac{3}{2}\sqrt{y}$. The value of x at the point of intersection is 3. What is the corresponding value of y?

June '02, #8: If x is a positive integer, $4x^{\frac{1}{2}}$ is equivalent to

    (1) $\dfrac{2}{x}$ (2) $2x$ (3) $4\sqrt{x}$ (4) $4\dfrac{1}{x}$

June '02, #14: What is the solution set of the equation $x = 2\sqrt{2x - 3}$?

    (1) { }   (2) {2}   (3) {6}   (4) {2,6}

Sample #6: Which expression is equivalent to $\dfrac{\sqrt{7} + \sqrt{2}}{\sqrt{7} - \sqrt{2}}$?

    (1) $\dfrac{9}{5}$ (2) -1 (3) $\dfrac{9 + 2\sqrt{14}}{5}$ (4) $\dfrac{11 + \sqrt{2}}{14}$

Aug '01, #4: The solution set of the equation $\sqrt{x + 6} = x$ is

    (1) {-2,3}   (2) {-2}   (3) {3}   (4) { }

Jan '02, #18: What is the domain of $h(x) = \sqrt{x^2 - 4x - 5}$?

    (1) $\{x | x \geq 1 \text{ or } x \leq -5\}$ (3) $\{x | -1 \leq x \leq 5\}$
    (2) $\{x | x \geq 5 \text{ or } x \leq -1\}$ (4) $\{x | -5 \leq x \leq 1\}$

<u>June '03, #5:</u> Which expression is equivalent to $\dfrac{4}{3+\sqrt{2}}$?

(1) $\dfrac{12+4\sqrt{2}}{7}$ (2) $\dfrac{12+4\sqrt{2}}{11}$ (3) $\dfrac{12-4\sqrt{2}}{7}$ (4) $\dfrac{12-4\sqrt{2}}{11}$

<u>Jan '03, #5:</u> What is the solution set of the equation $\sqrt{9x+10} = x$?

(1) {-1}   (2) {9}   (3) {10}   (4) {10,-1}

<u>Jan '03, #14:</u> If $f(x) = \dfrac{1}{\sqrt{2x-4}}$, the domain of $f(x)$ is

(1) $x = 2$ (2) $x < 2$ (3) $x \geq 2$ (4) $x > 2$

<u>Aug '03, #7:</u> Which expression is equal to $\dfrac{2+\sqrt{3}}{2-\sqrt{3}}$?

(1) $\dfrac{1-4\sqrt{3}}{7}$ (2) $\dfrac{7+4\sqrt{3}}{7}$ (3) $1-4\sqrt{3}$ (4) $7+4\sqrt{3}$

<u>Jan '04, #5:</u> The expression $\dfrac{7}{2-\sqrt{3}}$ is equivalent to

(1) $14-7\sqrt{3}$ (2) $14+7\sqrt{3}$ (3) $\dfrac{2+\sqrt{3}}{7}$ (4) $\dfrac{14+\sqrt{3}}{7}$

<u>Jan '04, #13:</u> The expression $b^{-\frac{3}{2}}, b>0$, is equivalent to

(1) $\dfrac{1}{(\sqrt[3]{b})^2}$ (2) $\dfrac{1}{(\sqrt{b})^3}$ (3) $-(\sqrt{b})^3$ (4) $(\sqrt[3]{b})^2$

<u>Jan '04, #27:</u> Solve algebraically: $\sqrt{x+5}+1 = x$

Review: Turning Points and Axis of Symmetry

Every parabola has either a maximum point or a minimum point, also known as its turning point.

With the calculator a close approximation of this point is easy to find.

a. Begin with the equation $y = x^2 + 7x + 6$.

b. Enter this equation in Y_1 and press GRAPH.

c. Decide whether the parabola has a lowest point or a highest point.

d. Press 2nd TRACE.

e. If the parabola has a lowest point choose 3:minimum. If it has a highest point choose 4:maximum.

f. When asked "Left Bound?", move the cursor to a point left of the point you are interested in and press ENTER.

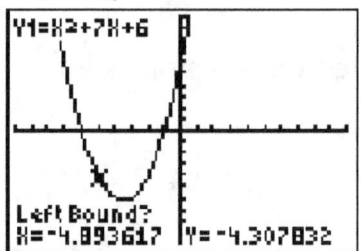

g. When asked "Right Bound?", move the cursor to a point right of the cursor and press ENTER.

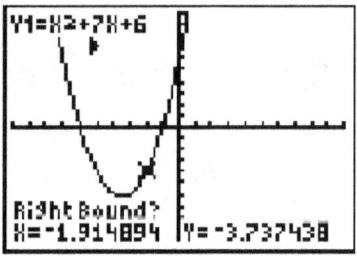

h. Move the cursor near the turning point as your "Guess" and press ENTER.

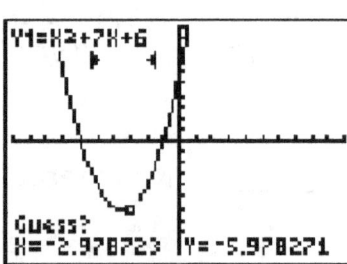

i. The *x* and *y* values at the bottom of the screen represent the coordinates of the maximum or minimum (it will tell you which one).

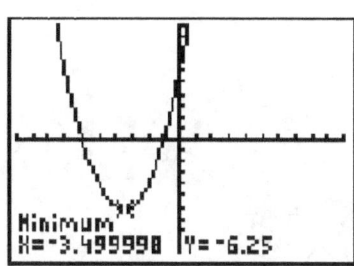

Note that in this example there are several 9's following the decimal value 4. The calculator will occasionally give these trailing 9's or trailing 0's when the actual value is the rounded value (it thought it was "close enough"). The turning point of this parabola is
(-3.5,-6.25)

j. If you need to identify the axis (or line) of symmetry, notice that the parabola has line symmetry and this line must pass through the turning point. Therefore, if you have found the maximum or minimum, you have found the *x*-value that this line must pass through. Since it must be a vertical line the equation of the line is
x=-3.5

Practice:

Find the maximum or minimum for each quadratic equation and give the equation of the axis of symmetry for the parabola it represents.

	Turning Point	Axis of Symmetry
1. $y = -x^2 + 5x + 3$	_____	_____
2. $y = 2x^2 + 5x - 9$	_____	_____
3. $y = 3x^2 - 2x - 8$	_____	_____
4. $y = -2x^2 + x + 4$	_____	_____
5. $y = x^2 - 6x + 9$	_____	_____
6. $y = -x^2 + 4x + 12$	_____	_____
7. $y = 4x^2 - 5x - 3$	_____	_____
8. $y = \frac{1}{2}x^2 - 2x - 3$	_____	_____

Graphing Quadratic Equations

The graph of a quadratic equation is called a _____.

The most important characteristics of a quadratic equation are _____ and _____ .

Frequently we will be asked to show the graph of a quadratic equation. When we are, we must identify these important characteristics.

Also, if it is part of a system (_____) we must show any points of intersection with the other equation (usually a line).

Below are the NYS guidelines for graphing and sketching. You will be responsible for following these guidelines on all graphs, including graphs of parabola.

"Sketches and graphs should contain the following:
- ➤ The labeled graph of each equation when more than one function is graphed (no deduction if only one of two is not labeled
- ➤ Axes appropriately labeled – variables identified
- ➤ Intercepts noted, where appropriate
- ➤ Points of intersection labeled
- ➤ Indicate window used by showing any of the following:
  - o Intercepts
  - o Scale on both axes
  - o Maximum and minimum values of x and y
- ➤ In the graphs of nonlinear functions at least three points should be indicated. Intercepts are acceptable, and when appropriate, the turning point should be indicated in the graph of the parabola
- ➤ If a student sketches a graph not on a grid for problems where grid use is optional, the above criteria for sketches and graphs still hold."

The first thing you should check is to see if there is a given or implied scale or window you should use.

Example:
1. The instant replay facility at the Superdome in New Orleans, Louisiana, was moved because a high punt kicked by Oakland Raider Ray Guy hit it. The original position of the facility was 90 feet above the playing field. Jason, a high school punter, can kick a football with an initial velocity of 65 feet per second. The height of the football t seconds after he kicks it is found by the function $h(t) = -16t^2 + 65t$.

Note: If you are having difficulty finding an appropriate viewing window, go on to the next section, Review of Zoom and Window, before proceeding with the graphs in this lesson.

     a. What is the implied minimum height? (minimum *y*-value) _____

     b. What is the implied maximum height? (maximum *y*-value) _____

     c. What are the *x*-intercepts? _____ and _____

     d. What is the turning point? _____

     e. Graph the parabola in an appropriate window and copy on the grid below.

     f. Label to meet NYS requirements.

If Jason were to have kicked a football in the Superdome before the instant replay facility was moved, would he have been able to hit it? Explain.

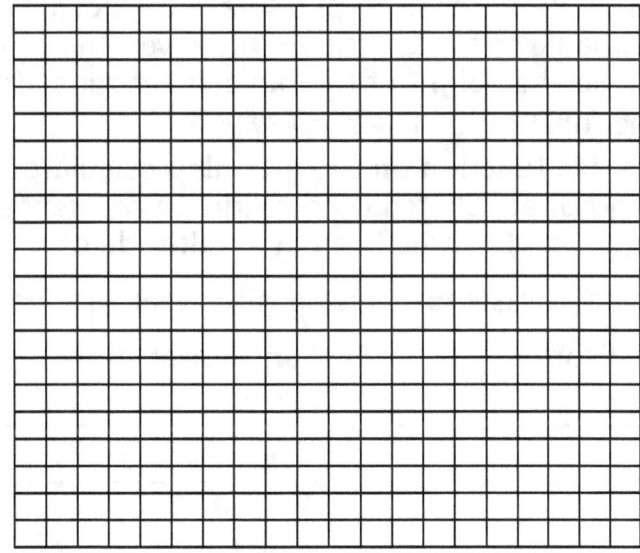

2. Show the solution by graphing:
The Empire State Building is 1250 feet tall. If an object is thrown upward from the top of the building at an initial velocity of 25 feet per second, its height t seconds after it is thrown is given by the function $h(t) = -16t^2 + 25t + 1250$.

   a. Sketch a graph that represents this function.

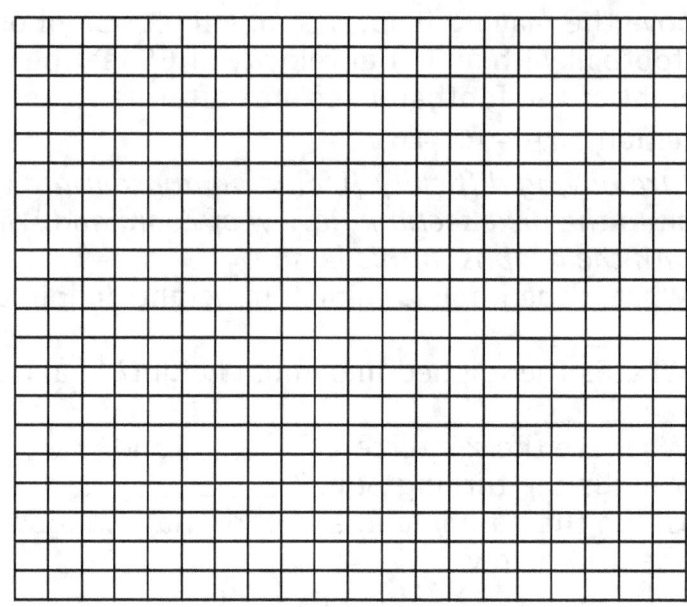

   b. Tell how long it will take for the object to hit the ground.

   _____

3. Police are investigating the shooting of a police helicopter. They found a weapon at the scene of the crime that has a suspect's fingerprints on it. Forensic experts have deduced that the weapon is capable of firing with an initial velocity of 890 feet per second. So the height of the bullet s seconds after firing is found by the function $h(s) = -25s^2 + 890s$.

   a. Graph this function on the graphing calculator and make an appropriate sketch on the grid *on the next page*.
   If the helicopter was flying at an altitude of 6500 feet at

   the time it was shot, is it possible that this weapon shot

   the helicopter? _____ Explain your answer.

   _____

   _____

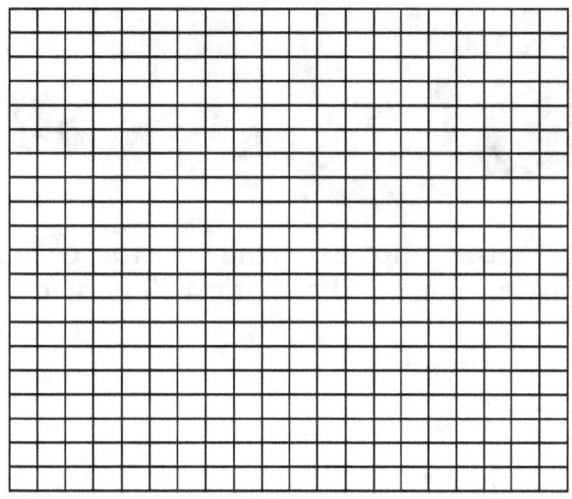

4. A student is tossing a stone upward on the school lawn. Its height y, in feet, is given by the equation $y = -16t^2 + 40t$ where t represents the number of seconds since the stone was thrown.

   a. Sketch the graph of the parabola below.

   b. There is a branch with a nest of baby birds 26 feet above the ground. Are the birds in danger of being his by the stone? ___ Explain.

   _____

   _____

   c. The principal is walking by 3 seconds after the stone is thrown. Will he be hit? _____ Explain.

   _____

   d. Write the equation for the axis of symmetry of this parabola.

   _____

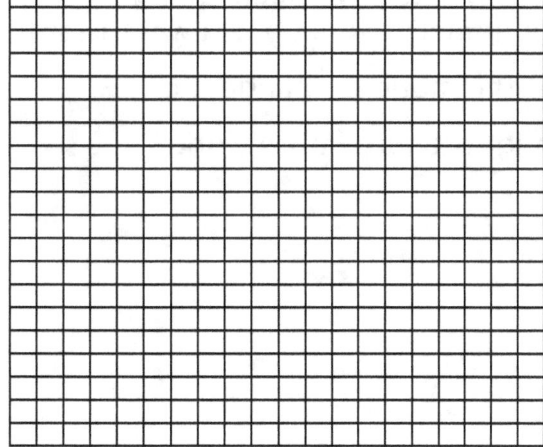

The standard window is not especially useful for Math B questions, but it is a good starting point. The window dimensions in the standard window are:

Xmin= _____

Xmax= _____

Xscl= _____

Ymin= _____

Ymax= _____

Yscl= _____

Xres= _____

These dimensions can give a distorted image because

_____

_____

To "fix" the distortion use _____ or _____ .
Either of these choices will

_____

When graphing statistics plots, always use _____ .
***Repeat whenever the lists have been changed.
When graphing trigonometric functions always use _____

***Repeat whenever the mode is changed. (_____)

Quadratic graphs might require more work.

Grids provided for regents questions are *usually* 20 squares by 20 squares.

The following method will work for most quadratic questions to find a "good" window:

1. Enter the equation in Y=.
2. Press ZOOM and choose 0:ZoomFit. This is not the window we would like but will help us quickly bring the critical points into the viewing window. (Note: If this doesn't work, try Zoom 6 first.)
3. The *x*-axis is usually okay left as it is, but if negative values do not make sense in the context of the question, change the Xmin to 0.
4. If the roots of the equation do not fall into the interval from 0 to 20, you will need to change the x-scale on your grid, but this is not usually necessary.
5. The Ymin will usually need to be changed to 0. Change the Ymax to the next multiple of 20 greater than the Ymax. Use this multiplication factor as the *y*-scale on your grid.

or

6. It is often quicker to use the formula for the axis of symmetry to find the maximum or minimum value for *y*. If you have not already memorized the formula, it is $\boxed{x = \dfrac{-b}{2a}}$. Double this value to find a "good" maximum *x*. Evaluate the equation at this *x*-value to find a "good" maximum or minimum *y*-value.

If errors occur when graphing:

The two most frequently occurring errors are ERR:INVALID DIM and ERR:DIM MISMATCH

The first error occurs when graphing statistics plots where two lists are necessary, but only one has data in.

The second error occurs when graphing statistics plots where two lists are necessary, but one is longer than the other; or when a stat plot has not been turned off and another type of graph is being attempted.

An error may also occur if you have made your minimum window value greater than or equal to the maximum value.

June '02, #9: What is the equation of a parabola that goes through the points (0,1), (-1,6), and (2,3)?

(1) $y = x^2 + 1$ (2) $y = 2x^2 + 1$ (3) $y = x^2 - 3x + 1$ (4) $y = 2x^2 - 3x + 1$

June '02, #25: The equation $W = 120I - 12I^2$ represents the power (*W*) in watts, of a 120-volt circuit having a resistance of 12 ohms when a current (*I*) is flowing through the circuit. What is the maximum power, in watts, that can be delivered in this circuit?

June '02, #28: A pelican flying in the air over water drops a crab from a height of 30 feet. The distance the crab is from the water as it falls can be represented by the function $h(t) = -16t^2 + 30$, where *t* is time, in seconds. To catch the crab as it falls, a gull flies along a path represented by the function $g(t) = -8t + 15$. Can the gull catch the crab before the crab hits the water? Justify your answer. [The use of the accompanying grid is optional.]

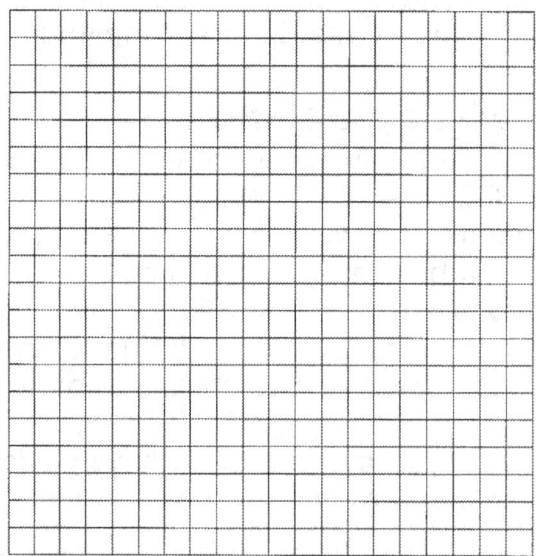

<u>Sample #15</u>: A model rocket is launched from ground level. Its height, h meters above the ground, is a function of time t seconds after launch and is given by the equation $h = -4.9t^2 + 68.6t$. What would be the maximum height, to the nearest meter, attained by the model?

(1) 243   (2) 242   (3) 241   (4) 240

<u>Aug '01, #12</u>: A ball is thrown straight up at an initial velocity of 54 feet per second. The height of the ball t seconds after it is thrown is given by the formula $h(t) = 54t - 12t^2$. How many seconds after the ball is thrown will it return to the ground?

(1) 9.2   (2) 6   (3) 4.5   (4) 4

<u>Jan '02, #31</u>: When a baseball is hit by a batter, the height of the ball, $h(t)$, at time t, $t \geq 0$, is determined by the equation $h(t) = -16t^2 + 64t + 4$. For which interval of time is the height of the ball greater than or equal to 52 feet?

<u>June '01, #1</u>: An archer shoots an arrow into the air such that its height at any time, t, is given by the function $h(t) = -16t^2 + kt + 3$. If the maximum height of the arrow occurs at time $t = 4$, what is the value of k?

(1) 128   (2) 64   (3) 8   (4) 4

<u>June '01, #28</u>: A homeowner wants to increase the size of a rectangular deck that now measures 15 feet by 20 feet, but the building code laws state that a homeowner cannot have a deck larger than 900 square feet. If the length and the width are to be increased by the same amount, find, to the nearest tenth, the maximum number of feet that the length of the deck may be increased in size legally.

<u>June '03, #21:</u> Vanessa throws a tennis ball in the air. The function $h(t) = -16t^2 + 45t + 7$ represents the distance, in feet, that the ball is from the ground at any time t. At what time, to the nearest tenth of a second, is the ball at its maximum height?

<u>Aug '02, #29:</u> A rock is thrown vertically from the ground with a velocity of 24 meters per second, and it reaches a height of $2 + 24t - 4.9t^2$ after t seconds. How many seconds after the rock is thrown will it reach maximum height, and what is the maximum height the rock will reach, in meters? How many seconds after the rock is thrown will it hit the ground? Round your answers to the nearest hundredth. [Only an algebraic or graphic solution will be accepted.]

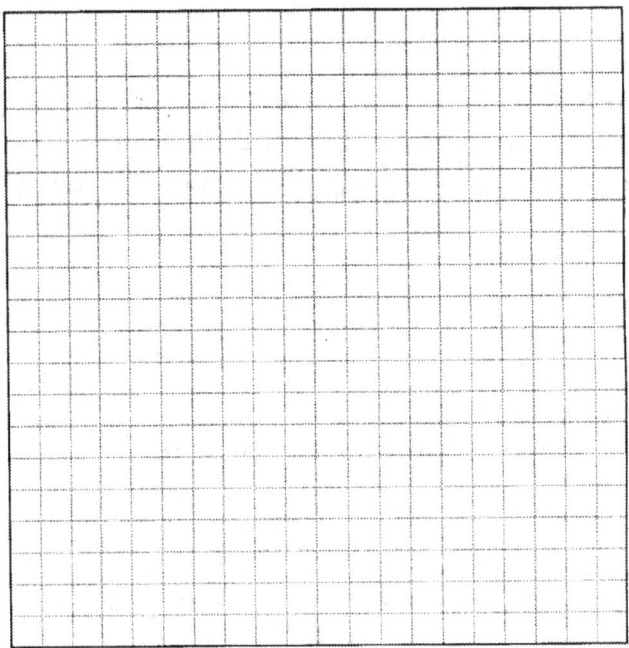

<u>Jan '03, #3:</u> Which equation represents the parabola shown in the accompanying graph?

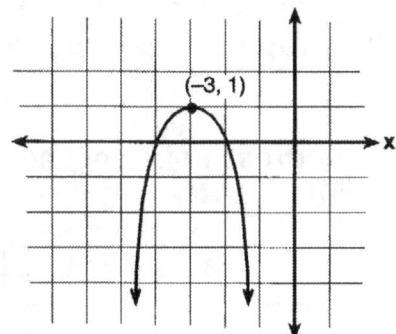

(1) $f(x) = (x+1)^2 - 3$ (3) $f(x) = -(x+3)^2 + 1$

(2) $f(x) = -(x-3)^2 + 1$ (4) $f(x) = -(x-3)^2 - 3$

<u>Jan '03, #18:</u> Which graph represents the solution set of $x^2 - x - 12 < 0$?

(1)
-4 -3 -2 -1 0 1 2 3 4

(2)
-4 -3 -2 -1 0 1 2 3 4

(3)
-4 -3 -2 -1 0 1 2 3 4

(4)
-4 -3 -2 -1 0 1 2 3 4

<u>Jan '03, #22:</u> The height of an object, *h(t)*, is determined by the formula $h(t) = -16t^2 + 256t$, where *t* is time, in seconds. Will the object reach a maximum or a minimum? Explain or show your reasoning.

<u>Aug '03, #21:</u> The height *h*, in feet, a ball will reach when thrown in the air is a function of time, *t*, in seconds, given by the equation $h(t) = -16t^2 + 30t + 6$. Find, to the *nearest tenth*, the maximum height, in feet, the ball will reach.

Jan '04, #31: An acorn falls from the branch of a tree to the ground 25 feet below. The distance S, the acorn is from the ground as it falls is represented by the equation $S(t) = -16t^2 + 25$, where t represents time, in seconds. Sketch a graph of this situation on the accompanying grid.

Calculate, to the nearest hundredth of a second, the time the acorn will take to reach the ground.

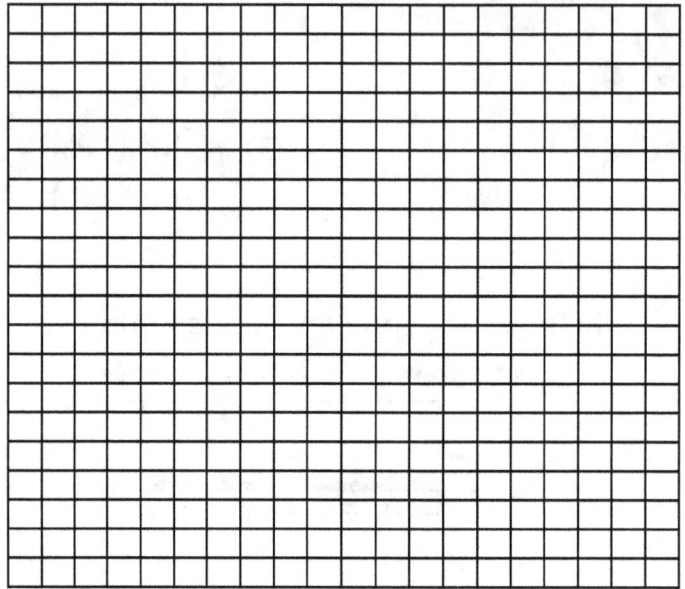

Jan '04, #24: When a current, I flows through a given electrical circuit, the power, W, of the circuit can be determined by the formula $W = 120I - 12I^2$. What amount of current, I, supplies the maximum power, W?

Graphing Absolute Value

In this lesson we will practice graphing absolute value functions and try to determine how the parts of the equation affect the finished product.

To graph absolute value functions we will need to remember how to find absolute value on the calculator. We can find it by pressing
_____, moving over to _____ , and choosing
_____ .
Or, if we really can't remember how to get there, we can resort to finding it in the _____ . (_____)

Let's set up a general equation so that we can refer to the parts of the equation more easily:

$$y = |ax + b| + c$$

To see what the most basic absolute value equation looks like, graph $y = |x|$ on the graphing calculator and sketch the screen in the box below.

```
Plot1  Plot2  Plot3
\Y1∎abs(X)
\Y2=
\Y3=
\Y4=
\Y5=
\Y6=
\Y7=
```

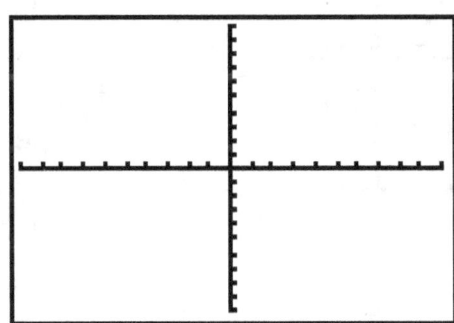

What do you notice about the values of *y*? _____
What makes this equation different than others you have graphed?
_____

In this lesson you may make "quick" sketches, but be sure the turning point of the equation is accurately placed!

Graph and sketch the following equations.

1. $y = |2x|$

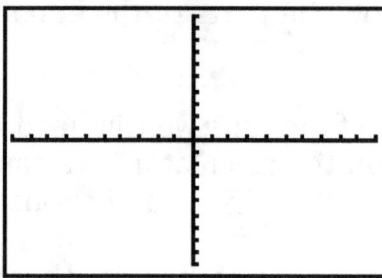

2. $y = |5x|$

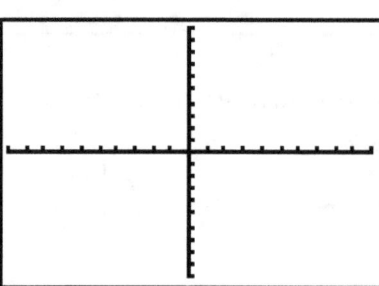

3. $y = |10x|$

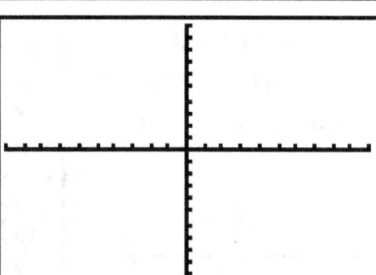

4. $y = \left|\frac{1}{2}x\right|$

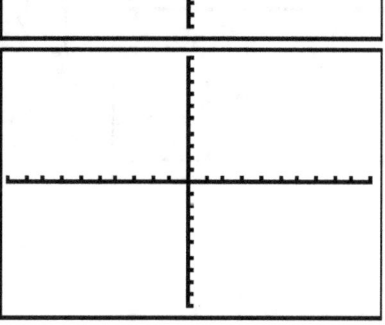

5. $y = |.1x|$

What conclusion can you make about the effect that "a" has on the appearance of the graph of the absolute value equation?

_____
_____
_____
_____

6. $y = |x-1|$

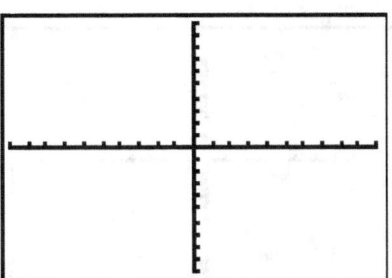

7. $y = |x+1|$

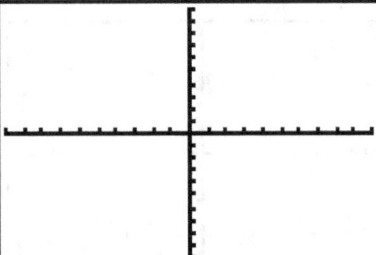

8. $y = |x-3|$

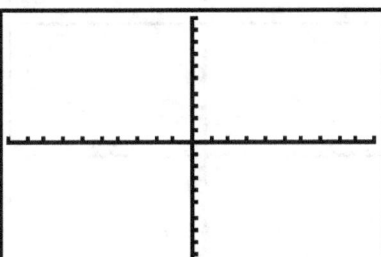

9. $y = |x+3|$

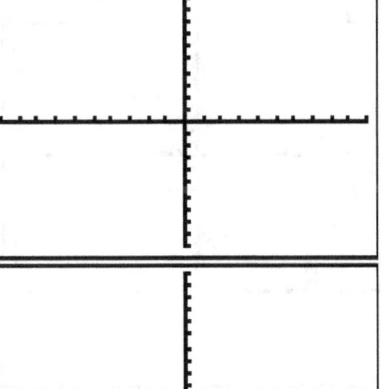

10. $y = |x-5|$

What conclusion can you make about the effect that "b" has on the appearance of the graph of the absolute value equation?

_____
_____
_____
_____

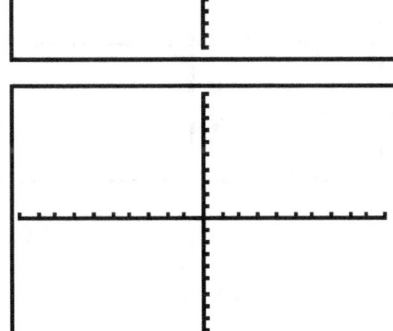

11. $y = |x+5|$

12. $y = |x| - 1$

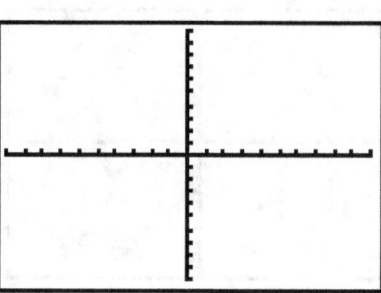

13. $y = |x| + 4$

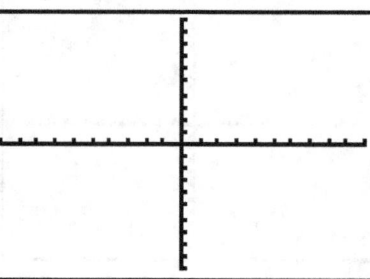

What conclusion can you make about the effect that "c" has on the appearance of the graph of the absolute value equation?

_____

_____

_____

_____

14. $y = |x| - 5$

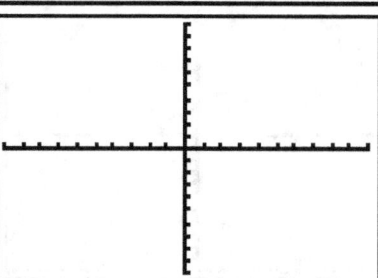

15. $y = |x| + 8$

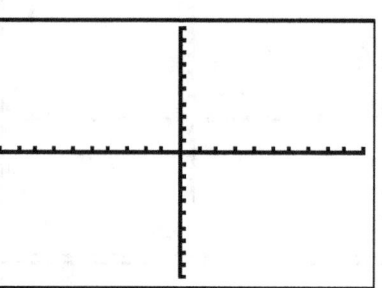

Try these.

16. $y = 3|x|$

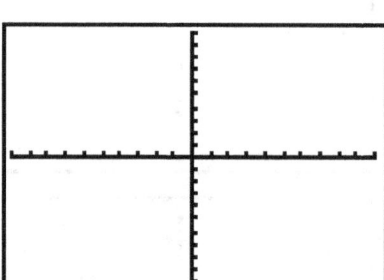

17. $y = -|x|$

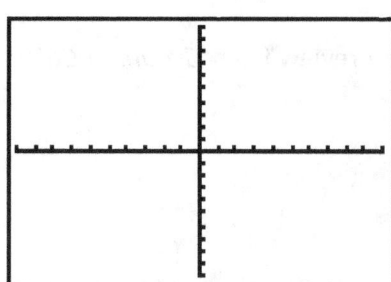

18. $y = -3|x + 1|$

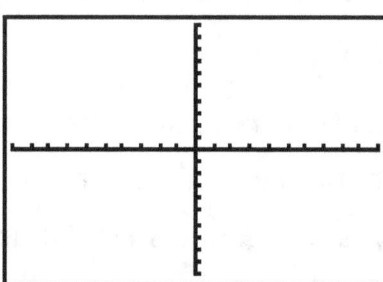

19. $y = -|2x + 5| + 3$

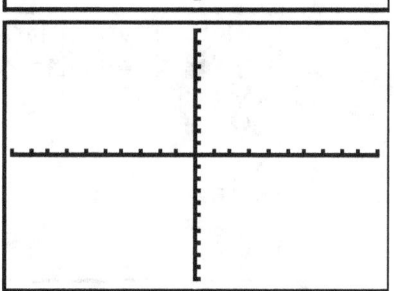

Can you find the equation that matches these graphs?

20.

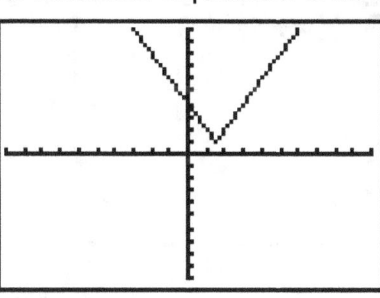

21.

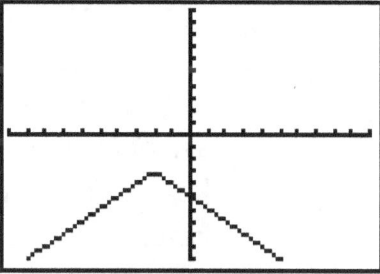

22.

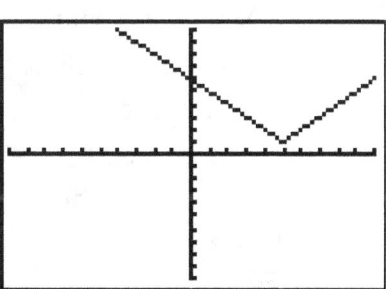

23.

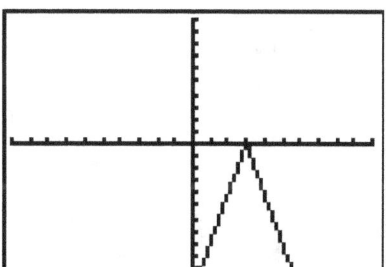

The following lessons are a review from *Graphing Calculator for NYS Math A and Beyond*.

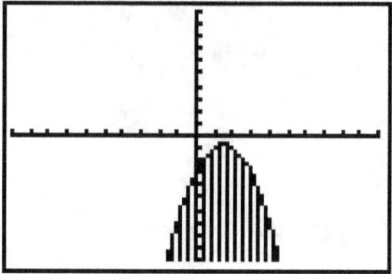

1. Enter the inequality in Y_1 as an equation.

2. Press the left arrow key until the line to the left of Y_1 is flashing.

3. If the inequality is < or ≤ press ENTER until this symbol appears:

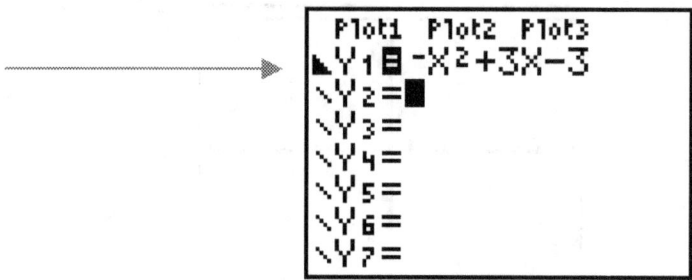

4. If the inequality is > or ≥ press ENTER until this symbol appears:

If you go past the symbol you need, keep going, it will return.

5. Press GRAPH.

Note that the calculator will not distinguish between < and ≤. You will need to know which one you need.

Practice:
Graph each inequality on the calculator.
Sketch the graph. Use a standard window.

a. $y > x^2 - x - 6$

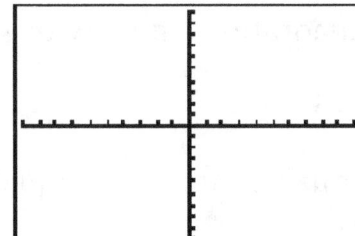

b. $y < 4x^2 - 9$

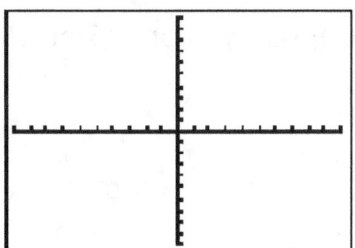

c. $y \leq x^2 + 2x - 15$

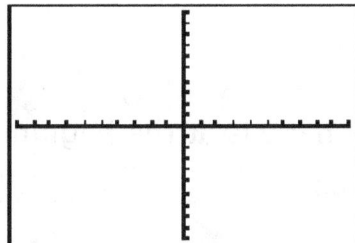

d. $y < -2x^2 + 16$

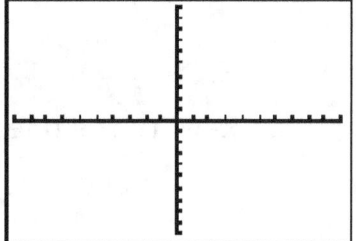

e. $y \geq -x^2 + 3x - 2$

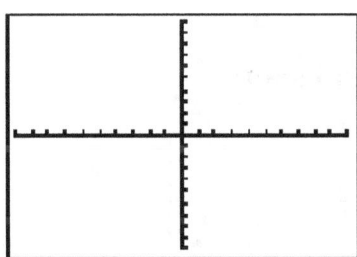

f. $y > -4x^2 + 25$

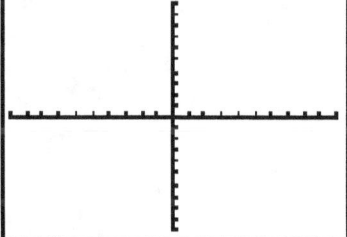

g. $y < \frac{1}{4}x^2 + x + 1$

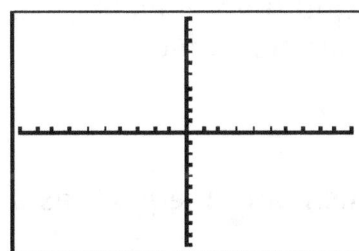

h. $y \leq \frac{4}{9}x^2 + 2x - 5$

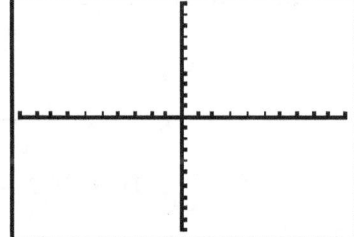

Using Systems to Solve Quadratic Inequalities

Suppose you are given the following quadratic inequality to solve:

$$x^2-2x+1<5$$

1. Begin by separating the inequality into two inequalities.

$$x^2-2x+1< \ldots\ldots\ldots <5$$

2. Place one y after the first < and another before the second.

$$x^2-2x+1<y \qquad y<5$$

3. Rewrite the one on the left so that y is at the beginning of the inequality. Remember, this will_____

_____   _____

4. Enter these in Y_1 and Y_2.

5. Graph and sketch.

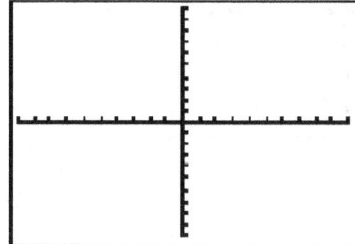

6. Use the intersect command to find the values for x at which the line and the parabola meet.

_____   _____

7. Describe the x values for which both inequalities are true.

_____

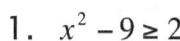

Practice:

a. Enter the inequality as two inequalities in Y_1 and Y_2.

b. Graph in a standard window and sketch the screen neatly.

c. Find the points of intersection.

d. Describe the values of x for which the original inequality are true.

1. $x^2 - 9 \geq 2$

a. & b.

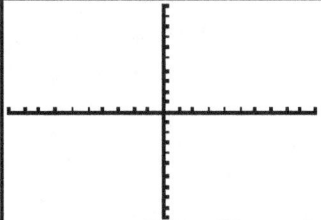

   c. _____   _____

   d. _____

2. $x^2 - 4x + 4 \leq 6$

a. & b.

   c. _____   _____

   d. _____

3. $2x^2 + 3x - 5 < -1$

a. & b.

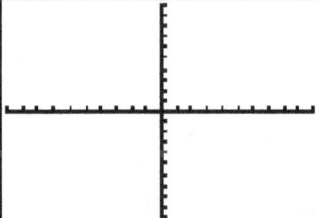

   c. _____   _____

   d. _____

4. $-x^2 + 25 \leq 4$

a. & b.

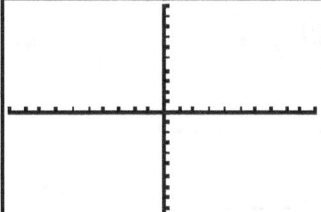

   c. _____   _____

   d. _____

5. $-x^2 + 6x - 9 \geq 0$

a. & b.

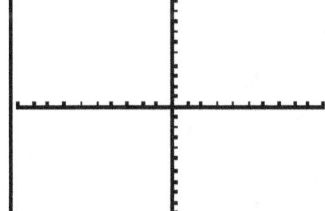

   c. _____   _____

   d. _____

The same method works for absolute value inequalities!

6. $|2x + 3| \geq 7$ a. & b.

     c. _____ _____

     d. _____

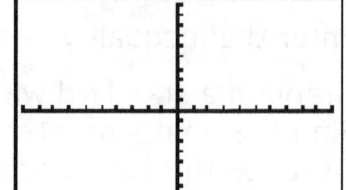

7. $|2 - x| > 8$ a. & b.

     c. _____ _____

     d. _____

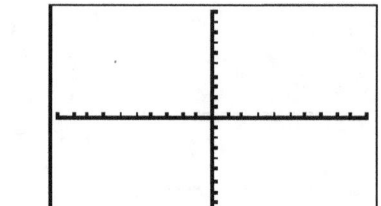

8. $|2 + 3x| \geq 4$ a. & b.

     c. _____ _____

     d. _____

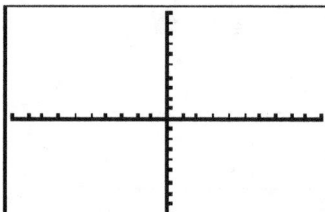

9. $|6 + x| \leq 1$ a. & b.

     c. _____ _____

     d. _____

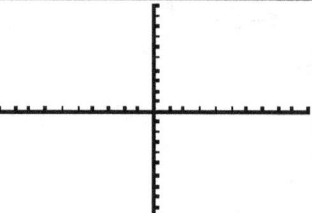

10. $|x + 4| < 2$ a. & b.

     c. _____ _____

     d. _____

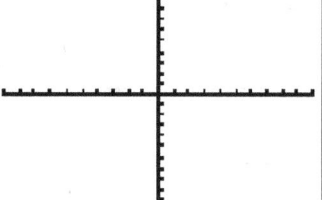

Absolute Value

<u>Aug '01, #2:</u> The solution set of |3x+2|<1 contains

    (1) only negative real numbers
    (2) only positive real numbers
    (3) both positive and negative real numbers
    (4) no real numbers

<u>June '01, #7:</u> Which equation states that the temperature, t, in a room is less than 3° from 68°?

    (1) |3-t|<68   (3) |68-t|<3
    (2) |3+t|<68   (4) |68+3|<t

<u>June '03, #18:</u> What is the solution set of the inequality |3-2x| ≥ 4?

    (1) $\{x|\frac{7}{2} \le x \le -\frac{1}{2}\}$ (3) $\{x|x \le -\frac{1}{2}\ \text{or}\ x \ge \frac{7}{2}\}$

    (2) $\{x|-\frac{1}{2} \le x \le \frac{7}{2}\}$ (4) $\{x|x \le \frac{7}{2}\ \text{or}\ x \ge -\frac{1}{2}\}$

<u>Aug '02, #3:</u> What is the solution of the inequality $|x+3| \le 5$?

    (1) −8 ≤ x ≤ 2   (2) −2 ≤ x ≤ 8   (3) x≤-8 or x≥2   (4) x≤-2 or x≥8

<u>Jan '03, #26:</u> The inequality $|1.5C - 24| \le 30$ represents the range of monthly average temperatures, C, in degrees Celsius, for Toledo, Ohio. Solve for C.

Aug '03, #3: Which graph represents the solution set of $|2x - 1| < 7$?

(1)

(2)

(3)

(4)

Jan '04, #14: The graph below represents f(x).

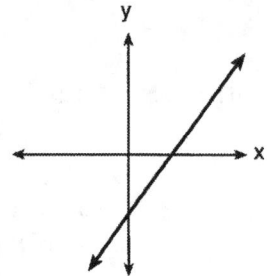

Which graph best represents $|f(x)|$?

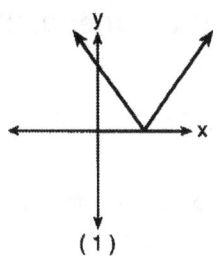

(1)

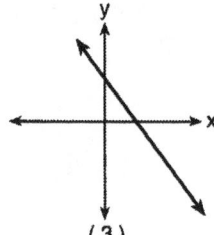

(3)

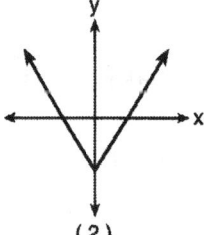

(2)

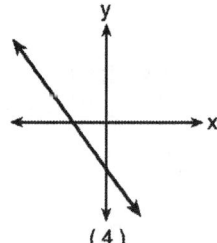

(4)

Entering Lists and 1-Variable Statistics

In order to use any choices from the statistics menu, you must begin with data. Although you may create your own list names, it is usually easier to use the list names L_1 - L_6 already on the calculator. Because these lessons will refer to these list names and it is possible to delete these from your calculator, we'll begin by telling you how to return these so that you can "stay on the same page".

If you have inadvertently deleted a list numbered 1-6, you can retrieve the list, including any data that was in the list by:

1. Pressing **STAT**
2. Choosing **5:SetUpEditor**
3. Press **ENTER** when the calculator takes you back to the home screen.

Lists 1-6 should now be restored.

We need data in our list before we can begin any statistics work. Let's consider the data below.

U.S. Greenhouse Gas Emissions From Human Activities

Year	Total Emissions	
1990	6139.6	(Measured in teragrams of carbon dioxide equivalents)
1995	6514.9	
1996	6707.0	
1997	6782.6	
1998	6801.3	
1999	6849.5	
2000	7047.4	
2001	6936.2	

Source: US Environmental Protection Agency

***Years are given for reference only; use total emissions for this exercise.

1. Press STAT.
2. Press ENTER or 1 to choose Edit.

3. Enter the data in List 1 (L_1). Be sure to press ENTER or the down arrow after each entry.

To find basic statistical information on your list, press STAT, go over to CALC, choose 1:1-Var Stats.
List the information you are given below:

Note the arrow beside n. This indicates that there is more information below.

What does all this mean?

Naming Lists

Sometimes we will want to save our lists for future use. The lists we will make for the following practice are long and we will be using them in later lessons. We also don't want to "tie up" lists 1-6.

To save work later, we will create new lists with names that will be stored in our calculators until we need them again.

***Note that if you are in the habit of resetting your calculator you

SHOULD NOT RESET until you are finished with these lists

or you will need to reenter them!!

1. Press STAT and choose 1:Edit.
2. Press the UP ARROW so that the list name is highlighted.
3. Use the RIGHT ARROW until you are one list past List 6. (Note that if you are not on the list name, the calculator will not allow you to go beyond the already named lists. Also, you cannot take the usual short cut. Going left from List 1 won't work here.)
4. There should be an A in the upper right corner meaning that the calculator has automatically changed to ALPHA mode.
5. Use up to five letters and/or numbers to name your list.
6. Enter your list as usual.

To retrieve this list you will need to:
1. Press 2nd
2. STAT (LIST)
3. Choose your list. (After the lists 1-6 the remaining lists should be in alphabetical order.)
4. Press ENTER.

Use the names, **PART, SULF,** and **NITRO** to enter the lists in the practice on the next page. These will remain on your calculator for you to use at a later time unless you reset either the RAM or ALL. If you choose to use other names you will need to remember them as the names above **may be referred to in later lessons or review assignments**.

If you forget and enter the lists in lists 1-6 you do not need to reenter them. Simply create the list as described above then, with the list name highlighted, enter the list name where the data is located and press ENTER.

The list should copy itself to its new location and you can clear the original list.

Practice:

Air Pollution in Selected Cities:

City and Country or State	Particulate Matter	Sulfur Dioxide	Nitrogen Dioxide
Amsterdam, Netherlands	37	10	58
Athens, Greece	50	34	64
Bangkok, Thailand	82	11	23
Barcelona, Spain	43	11	43
Beijing, China	106	90	122
Berlin, Germany	25	18	26
Cape Town, South Africa	15	21	72
Caracas, Venezuela	18	33	57
Chicago, Illinois	27	14	57
Delhi, India	187	24	41
Kolkata, India	153	49	34
London, UK	23	25	77
Los Angeles, California	38	9	74
Mexico City, Mexico	69	74	130
Milan, Italy	36	31	248
Montreal, Canada	22	10	42
Mumbai, India	79	33	39
New York, New York	23	26	79
Oslo, Norway	23	8	43
Paris, France	15	14	57
Prague, Czech Republic	27	14	33
Seoul, South Korea	45	44	60
Sofia, Bulgaria	83	39	122
Sydney, Australia	22	28	81
Tokyo, Japan	43	18	68
Toronto, Canada	26	17	43
Warsaw, Poland	49	16	32

Source: 2004 World Almanac

1. Find the following statistical measures for Particulate Matter:
   a. Mean: _____
   b. Median: _____
   c. Maximum: _____
   d. Minimum: _____
   e. Range: _____
   f. 25th Percentile: _____
   g. 75th Percentile: _____

2. Find the following statistical measures for Sulfur Dioxide:
   a. Mean: _____
   b. Median: _____
   c. Maximum: _____
   d. Minimum: _____
   e. Range: _____
   f. 25th Percentile: _____
   g. 75th Percentile: _____

3. Which country or state has an amount of Nitrogen Dioxide closest to the Mean in amount?

   _____

4. Which country or state has an amount of Nitrogen Dioxide closest to the Median amount? _____

5. Which country or state has the greatest amount of Nitrogen Dioxide?

   _____

6. Which country or state has the least amount of Nitrogen Dioxide?

   _____

7. Which country or state has an amount of Nitrogen Dioxide closest to the 25th percentile? _____

8. Which countries and/or states have amounts of Nitrogen Dioxide above the 75th percentile?

```
-----------------------
-----------------------
-----------------------
-----------------------
-----------------------
-----------------------
-----------------------
```

9. Using the Internet or other sources, find the major sources for particulate matter in air pollution.

```
-------------------------------------------------------------
-------------------------------------------------------------
-------------------------------------------------------------
```

10. Using the Internet or other sources, find the major sources for sulfur dioxide in air pollution.

```
-------------------------------------------------------------
-------------------------------------------------------------
-------------------------------------------------------------
```

11. Using the Internet or other sources, find the major sources for nitrogen dioxide in air pollution.

```
-------------------------------------------------------------
-------------------------------------------------------------
-------------------------------------------------------------
```

If 1-variable statistics (1:1-Var Stats) gives us basic statistical data, what does 2-variable statistics do?

2-Variable Statistics:

2-variable statistics allows us to compare two sets of data without calculating the measures separately.

Consider the data below:

Public Debt of the U.S.:

Fiscal Year	Debt (billion$)	Debt per Capita ($)	Interest Paid (billion$)
1983	1377.2	5870	128.8
1984	1572.3	6640	153.8
1985	1823.1	7598	178.9
1986	2125.3	8774	190.2
1987	2350.3	9615	195.4
1988	2602.3	10534	214.1
1989	2857.4	11545	240.9
1990	3233.3	13000	264.8
1991	3665.3	14436	285.5
1992	4064.6	15846	292.3
1993	4411.5	17105	292.5
1994	4692.8	18025	296.3
1995	4974.0	18930	332.4
1996	5224.8	19805	344.0
1997	5413.1	20026	355.8
1998	5526.2	20443	363.8
1999	5656.3	20746	353.5
2000	5674.2	20591	362.0
2001	5807.5	20353	359.5
2002	6228.2	21589	332.5
2003	6783.2	23230	318.1

Source: 2004 World Almanac

1. Create list names DEBT, PCAP, and INT. Enter the data from the table with debt in billions of dollars in DEBT, debt per capita in dollars in PCAP, and interest paid in billions of dollars in INT.
2. Press STAT.
3. Go over to CALC.
4. Choose 2:2-Var Stats
5. The calculator will now be at the home screen. Recall that you will need to press 2nd, STAT and choose your list from the NAMES. First we will compare total debt with per capita debt so include DEBT,PCAP after 2-Var Stats and ENTER. (Be sure to include the comma between list names!)

List the results below:

$\bar{x} =$ ----------

$\Sigma x =$ ----------

$\Sigma x^2 =$ ----------

$Sx =$ ----------

$\sigma x =$ ----------

$n =$ ----------

$\bar{y} =$ ----------

$\Sigma y =$ ----------

$\Sigma y^2 =$ ----------

$Sy =$ ----------

$\sigma y =$ ----------

$\Sigma xy =$ ----------

$\min X =$ ----------

$\max X =$ ----------

$\min Y =$ ----------

$\max Y =$ ----------

What statistical measures are missing that were calculated for 1-variable statistics? _____

Using 2-variable statistics:

1. Compare total debt with interest paid. What are the mean amounts?

   _____ _____

2. What is the sum of the product of the debt and interest paid?

   _____

3. What is the range of interest paid?   _____

4. What is the range of total debt?   _____

Using 1-variable statistics:

5. Find the median amount of interest paid.   _____

6. Find the median amount of total debt.   _____

7. Find the median amount of per capita debt.   _____

8. Explain how the total debt can continue to increase while the interest paid begins to decrease.

   _____

   _____

9. Using the information in the table, find the approximate population of the U.S. in the years 1983 and 2003.

   _____ _____

A measure of dispersion shows how the data is spread out.

Box-and-Whisker plots are good pictures of dispersion.

Easy Box-and-Whisker Plots:

1. Enter data in L_1.

> Quiz Scores: 78, 90, 85, 72, 100, 84, 60, 85, 95, 55, 75, 88, 97, 74, 77, 80, 92, 69, 75, 83

2. Press STAT

3. Move over to CALC

4. Choose 1:1-Var Stats

5. ENTER (Unless you have more than one list. If there is more than one list you need to name the list you need the statistics for before pressing ENTER.)

6. If you need the mean it is the first entry on the list.

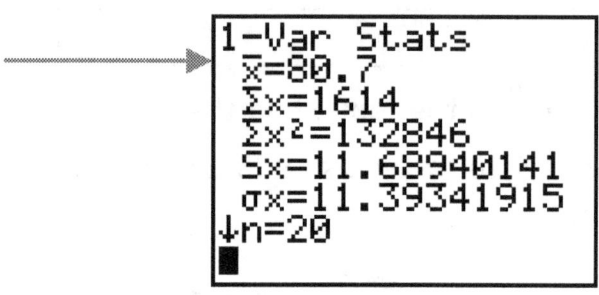

```
1-Var Stats
 x̄=80.7
 Σx=1614
 Σx²=132846
 Sx=11.68940141
 σx=11.39341915
↓n=20
```

7. Otherwise, scroll down until you see:

Minimum*
First (or Lower) Quartile*
Median*
Third (or Upper) Quartile*
Maximum*

```
1-Var Stats
↑n=20
 minX=55
 Q₁=74.5
 Med=81.5
 Q₃=89
 maxX=100
```

8. **The last five numbers are all you need and they are in the order that you need them.**
   Plot the minimum, first quartile, median, third quartile, and maximum just above the number line.

9. To find appropriate intervals, find the range of the data and divide by the number of intervals. Round up to a "nice" number. (i.e. a multiple of 2, 5, or 10 is usually easy to work with.)

10. Make vertical lines at the middle three points then make them into a box by adding the horizontal connecting lines.

11. Make "whiskers" extending out to the min and max.

12. To check your work you can see what the plot should look like by creating a STAT PLOT.

   a. Press 2ⁿᵈ , Y=.

   b. Turn on Plot 1 (Pressing ENTER twice should do it unless it is already turned on.)

   c. Use the arrow keys (down then left or right) to highlight the box-and-whisker plot. Be sure to choose the one with the vertical line inside the box at the median. Press ENTER.

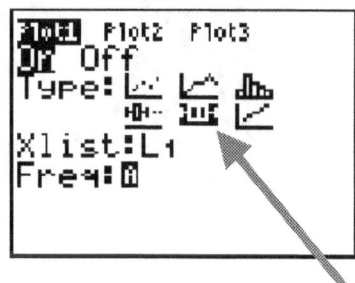

d. Be sure the Xlist matches the list name where your data was entered.
e. Press ZOOM and choose 9:ZoomStat. This will automatically fit your data to the type of plot you have chosen. ***If an INVALID DIM Error occurs the list you have entered as the Xlist probably has no data. Check your list name.

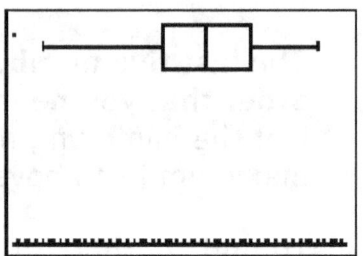

Practice:

Create box-and-whisker plots on the next page using the data at the right and answer the questions that follow.

Selected States	Total Population in thousands	Total Participants in Wildlife Associated Recreation in thousands
Maine	966	511
Vermont	455	242
New Hampshire	887	448
Connecticut	2514	928
Massachusetts	4726	1835
Rhode Island	759	284
New York	13944	3800
Pennsylvania	9298	3886
New Jersey	6129	1864
Ohio	8522	6281
Virginia	5168	2278
West Virginia	1467	593
Maryland	3912	1537
Delaware	560	232

Source: U.S. Fish and Wildlife Service 1996 National Survey of Fishing Hunting, and Wildlife-Associated Recreation

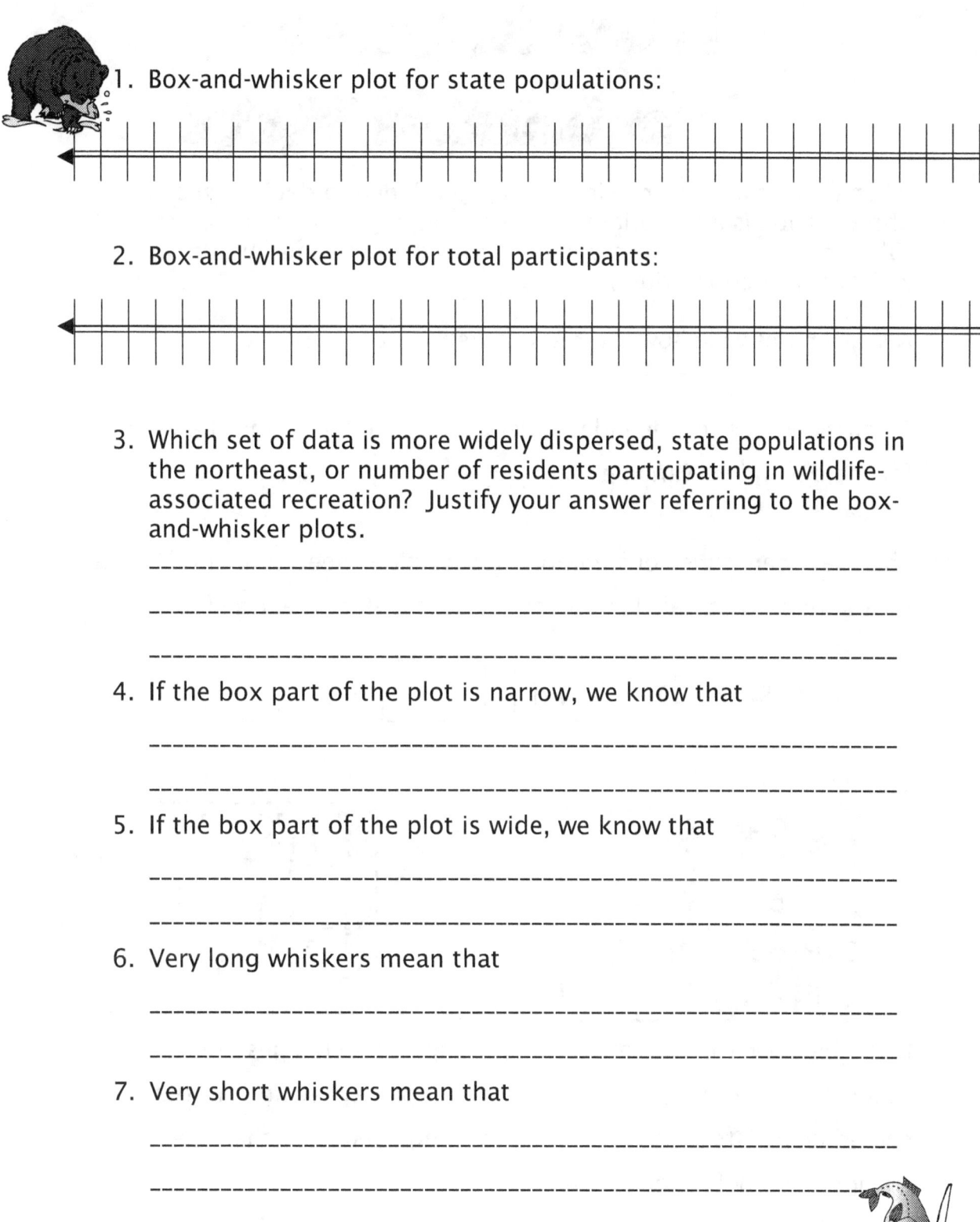

1. Box-and-whisker plot for state populations:

2. Box-and-whisker plot for total participants:

3. Which set of data is more widely dispersed, state populations in the northeast, or number of residents participating in wildlife-associated recreation? Justify your answer referring to the box-and-whisker plots.

    ----

4. If the box part of the plot is narrow, we know that

    ----

5. If the box part of the plot is wide, we know that

    ----

6. Very long whiskers mean that

    ----

7. Very short whiskers mean that

A Quick Course in Scatter Plots

When we calculate regressions it is a good idea to decide which type of regression is appropriate before calculating. (The alternative is to calculate each of the several types of regressions and find the one with the best correlation.)

A scatter plot will show if there is a trend in the data.

Enter the values for the independent variable (x) in L_1 and the values for the dependent variable (y) in L_2.

This may seem time consuming, but the values need to be in the lists for the regression calculation anyway so it is not an extra step.

Next, go to STATPLOT: $\boxed{\text{2nd}}$ $\boxed{\text{Y=}}$

Turn Plot 1 on:

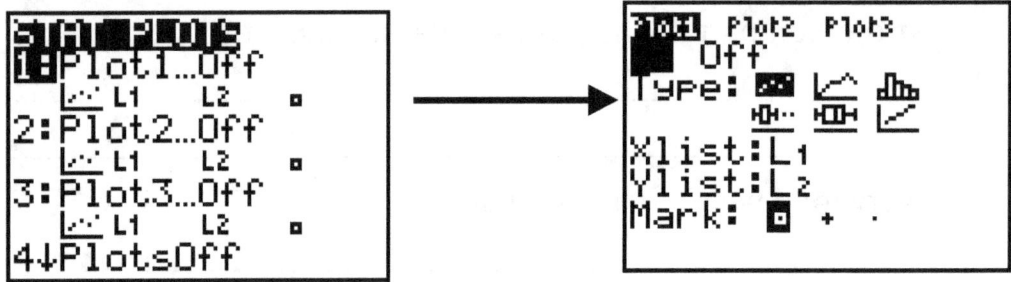

If you began with your calculator reset no other changes are necessary. The first picture is the scatter plot type and your data should be in lists 1 and 2. If you are using lists other than these change them before going on.

Press $\boxed{\text{Zoom}}$ and choose 9:ZoomStat

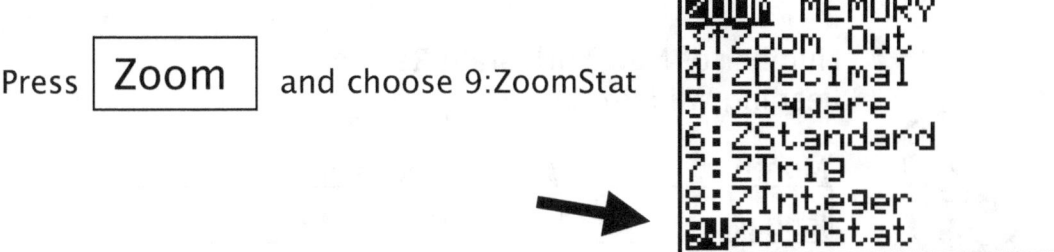

The calculator will choose the best viewing window based on the type of plot and the data in the lists.

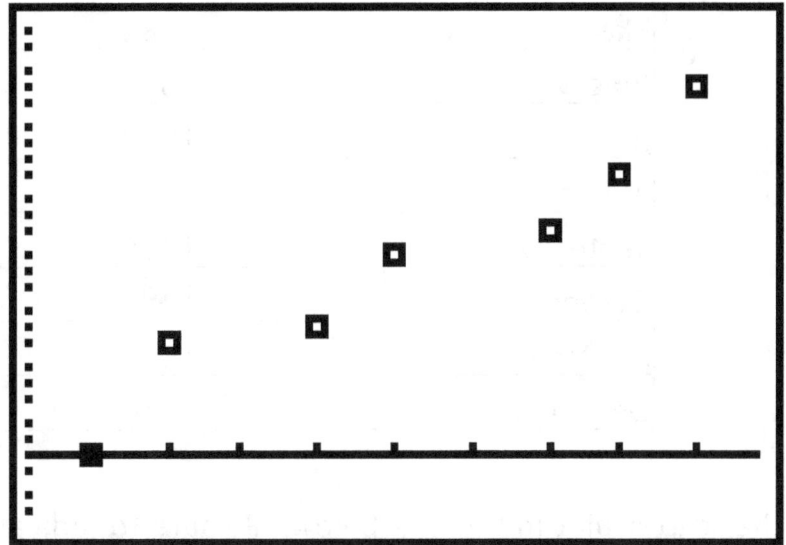

Look for a trend in the data points and choose the type of regression that will produce that line or curve.

When sketching your scatter plot be sure to include your window dimensions!!

You can find these by pressing WINDOW.
Make a scatter plot on the graphing calculator for the following set of data below. Use the number for the month as the independent variable (L_1) and the number of residents feeding birds as the dependent variable (L_2).

Number of New York State Residents Feeding Wild Birds (based on 1996 report)

Month	# of participants in thousands
January	2142
February	2160
March	2112
April	1982
May	1688
June	1666
July	1602
August	1514
September	1465
October	1532
November	1597
December	1716

Sketch the scatter plot in the box below. Be sure to indicate window dimensions.

Correlation

Correlation is an indication of relationships. A scatter plot shows us if there is a trend or a relationship. We will soon be computing a regression to find a rule to describe the relationship.

When executing a regression we will also want to know how well the data "fits" the equation. To do this we must activate the *correlation coefficient.*

Press ⬚ 2nd ⬚ then ⬚ 0 ⬚ to get the CATALOG. Go down to DiagnosticOn.

Press ⬚ ENTER ⬚ two times.

This is one of the few functions of the calculator that can be accessed only in the catalog!

The diagnostic will stay on until you turn it off (which you never NEED to do) or the calculator is reset.

(Leave the diagnostics on for the next part of the lesson.)

This will give us an "r" value that will tell us how well our data fits the equation found.

"r" Values

If all the data points lie exactly on the line and the slope of the line is positive (the line goes "uphill"), the correlation coefficient, r, is +1. We say that the data in the lists have a _____ _____ .

Similarly, if the data points are all exactly on the line and the slope of the line is negative (the line goes "downhill"), the correlation coefficient is –1. Now the data have a _____ _____ .

If, when you make your scatter plot, the points don't seem to make a line at all you will find that r≈0. In other words, if r is close to zero when you calculate the regression on the graphing calculator, it is not a "good fit".

How close it needs to be will depend on the type of data we are using:

1.   If the data is from a scientific experiment we would like to see

$$0.9 \leq r \leq 1 \quad \text{or} \quad -1 \leq r \leq -0.9$$

2.   If the data is from a social science experiment we would like to see

$$0.7 \leq r \leq 1 \quad \text{or} \quad -1 \leq r \leq -0.7$$

On the regents you will be told which type of regression to use regardless of whether it is the best type of regression or is truly the best fit. If it says **"line of best fit"** assume that you are to use a linear regression, but be prepared to discuss if it is a good fit or not.

You will also see "r^2". We will <u>not</u> need this value, <u>only the "r"</u>.

Also, not every regression has an appropriate r-value. You will typically only be asked for the correlation coefficient on linear regression questions.

Regression
Regression

The word *regression* was first used by Sir Francis Galton (1822-1911) in the late 19th century.

A study by Galton revealed that the height of the children of tall parents was above average, but seemed to fall back, *or regress*, toward the mean (average) height of the population.

Thus, the general process of predicting one variable (such as a child's height) based on another variable (the parent's heights) became known as regression.

The TI-83+/Ti-84+ will calculate many types of regression:

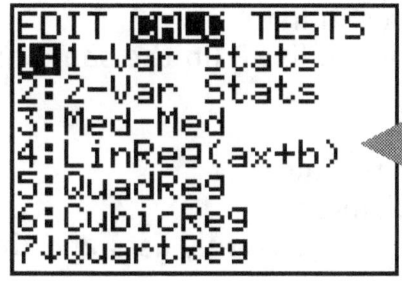

The most used type of regression will be 4:LinReg(ax+b). This is

a _____ _____and is also known as the

_____ _____ _____ _____ .

We will look at other types of regression later.

The above menu can be found by pressing $\boxed{\text{STAT}}$ and using the arrow key to move the cursor to CALC.

The linear regression requires that you fill two lists with data and that the lists be of equal length. The data should be matched in the lists and there should be two or more items in each list.
The first list should contain the _____variable. (*x*)

The second list should contain the _____variable. (*y*)

Think of it as: "As "*x*" changes, what happens to "*y*"?"
Or, in the case at the beginning, the parent's height was the independent variable. We already know how tall they were, but did the child's height depend on the parent's height?

Notes:
1. "a" is the _____ (what we usually call ___)
2. "b" is the _____
3. If the question does not specifically state the appropriate place to round to, write all of the digits visible on your screen!

Practice:
A. Starting really simple: find the equation of the line that passes through the points (6,-3) and (5,7).
 1. Enter the *x*'s in list 1.
 2. Enter the *y*'s in list 2 (in the same order the *x*'s were entered.)
 3. Press STAT.
 4. Move over to Calc
 5. Choose 4:LinReg
 6. If L_1 and L_2 are your only lists and they are in the correct order, press ENTER.
 7. If you have other lists or need to change the order, enter the list names with a comma between them then press ENTER.
 8. Write the equation with the given values of a _____ and b _____ in their correct places.
 9. The equation of the line through the given points is

 _____ .
 10.   What is the correlation coefficient for this regression equation? _____ Is this an unusual r-value?

 _____

 Explain:

 _____

 _____

B. On the Fahrenheit temperature scale, water freezes at 32° and boils at 212°; the corresponding temperatures on the Celsius scale are 0° and 100°. Given that the Fahrenheit and Celsius temperatures are linearly related, first find numbers a and b so that $F = aC + b$, and then answer these questions.

    1. Mercury freezes at -39°C. What is the corresponding

       Fahrenheit temperature?   _____

    2. For what value of C is F=0?   _____

    3. What temperature is the same in both scales?   _____

C. The boiling point of water is a function of altitude. The table shows the boiling points at different altitudes.

Location	Altitude h meters	Boiling Point of water t°C
Halifax, NS	0	100
Banff, Alberta	1383	95
Quito, Ecuador	2850	90
Mt. Logan	5951	80

1. Find the regression equation for the function.

    _____

2. Give the "r" value for this equation.

    _____

3. Discuss what this correlation coefficient means.

    _____

    _____

    _____

    _____

Stopping Distance

Speed	Thinking	Braking	Total Stopping Distance
20 mph	1.5 car lengths	1.5 car lengths	3 car lengths
30 mph	2.5 car lengths	3.5 car lengths	6 car lengths
40 mph	3 car lengths	6 car lengths	9 car lengths
50 mph	3.5 car lengths	12.5 car lengths	13 car lengths
60 mph	4.5 car lengths	13.5 car lengths	18 car lengths
70 mph	5 car lengths	19 car lengths	24 car lengths

Source:
Buckinghamshire County Council http://www.buckscc.gov.uk/road_safety/stopping-distance.htm

1. Using the table above, create a scatter plot for Thinking, Braking, and Total Stopping distances.
   Plot all three on the same screen using ■ for *Thinking*, + for *Braking*, and • for *Total Stopping* to distinguish the plots from one another.
   a. Do the relationships appear to be linear?_____
   b. Which set of data appears to create the straightest line?
   _____

2. Calculate the line of best fit for each set of data. Write the equation and the correlation coefficient.

   Thinking: _____   _____

   Braking: _____   _____

   Stopping: _____   _____

   Are the correlations "good"? Explain.

   _____

3. What appears to be the relationship between the correlation coefficients?
   _____

_____

_____

4. Why can none of these lines truly represent the situation?

    _____

    _____

    _____

5. Using the equation for Total Stopping Distance, find the length of time needed to stop for a car traveling each of the following speeds:

    55 mph: _____

    80 mph: _____

    100 mph: _____

    120 mph: _____

6. A tree has fallen across the road approximately 16-car lengths from a blind curve. What is the maximum speed at which a car could be traveling around this curve and successfully stop before hitting the tree?

    _____

7. The Department of Transportation needs to place a sign warning motorists of a Stop sign over the crest of a hill. The sign is visible at 10 car lengths from the intersection. What should the recommended speed on the sign be? Assume that many motorists will travel at approximately 5 mph above the recommended speed.

    _____

8. You are on your way to Watertown when you encounter a fast moving lake effect storm. Visibility is very poor. You estimate that you can see about two car lengths ahead. At what speed should you travel to avoid a possible collision? _____ It begins to clear a little and you can now see approximately 5 car lengths ahead. To what speed could you safely increase?

    _____

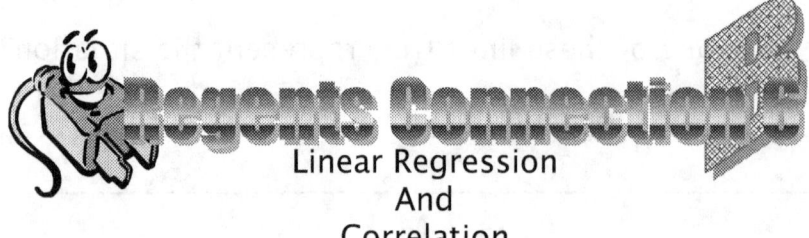

<u>June '01, #9:</u> The relationship of a woman's shoe size and length of a woman's foot, in inches, is given in the accompanying table,

Woman's Shoe Size	5	6	7	8
Foot Length (in)	9.00	9.25	9.50	9.75

The linear correlation coefficient for this relationship is

   (1) 1   (2) –1   (3) 0.5   (4) 0

<u>June '01, #34:</u> The 1999 win-loss statistics for the American League East baseball teams on a particular date is shown in the accompanying chart.

	W	L
New York	52	34
Boston	49	39
Toronto	47	43
Tampa Bay	39	49
Baltimore	36	51

Find the mean for the number of wins, $\overline{W}$, and the mean for the number of losses, $\overline{L}$, and determine if the point $(\overline{W}, \overline{L})$ is a point on the line of best fit. Justify your answer.

$\overline{W}$ = _____   Line of Best Fit: _____

$\overline{L}$ = _____

<u>Aug '01, #33:</u> The availability of leaded gasoline in New York State is decreasing, as shown in the accompanying table.

Year	1984	1988	1992	1996	2000
Gallons Available (in thousands)	150	124	104	76	50

Determine a linear relationship for *x* (years) versus *y* (gallons available), based on the data given. The data should be entered using the year and gallons available (in thousands), such as (1984,150).

If this relationship continues, determine the number of gallons of leaded gasoline available in New York State in the year 2005.

If this relationship continues, during what year will leaded gasoline first become unavailable in New York State?

<u>June '02, #11:</u> A linear regression equation of best fit between a student's attendance and the degree of success in school is *h=0.5x+68.5*. The correlation coefficient, r, for these data would be

(1) 0<r<1　　　　　(3) r=0
(2) –1<r<0　　　　　(4) r=-1

<u>Aug '03, #6:</u> Which graph represents data used in a linear regression that produces a correlation coefficient closest to –1?

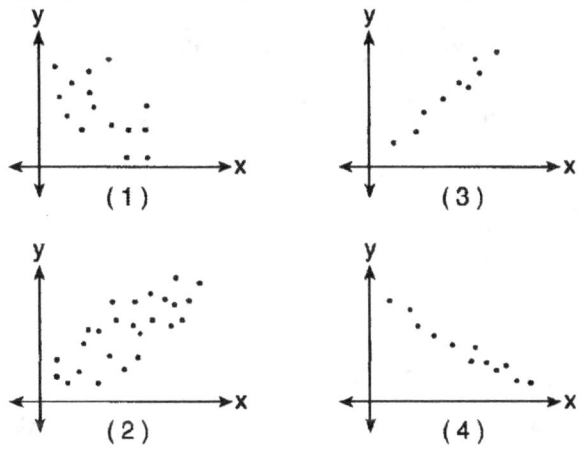

Jan '02, #34: Two different tests were designed to measure understanding of a topic. The two tests were given to ten students with the following results:

Test x	75	78	88	92	95	67	58	72	74	81
Test y	81	73	85	88	89	73	66	75	70	78

Construct a scatter plot for these scores, and then write an equation for the line of best fit (round slope and intercept to the nearest hundredth).

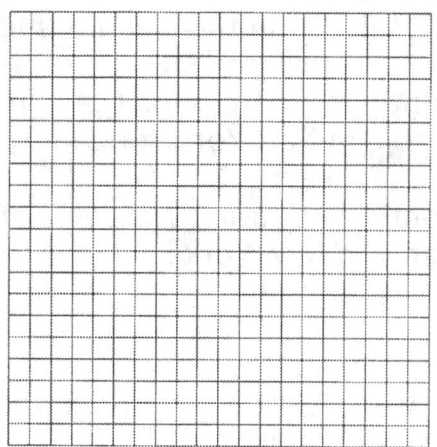

Find the correlation coefficient.
Predict the score, to the nearest integer, on test y for a student who scored 87 on test x.

Aug '03, #31: The table below shows the results of an experiment that relates the height at which a ball is dropped, x, to the height of its first bounce, y.

Drop Height (x) (cm)	Bounce Height (y) – (cm)
100	26
90	23
80	21
70	18
60	16

Find $\overline{x}$, the mean of the drop heights.
Find $\overline{y}$, the mean of the bounce heights.
Find the linear regression equation that best fits the data.
Show that $(\overline{x}, \overline{y})$ is a point on the line of regression. [Use of a grid is optional.]

<u>Jan '03, #28:</u> In a mathematics class of ten students, the teacher wanted to determine how a homework grade influenced a student's performance on the subsequent test. The homework grade and subsequent test grade for each student are given in the accompanying table.

Homework Grade	Test Grade
(x)	(y)
94	98
95	94
92	95
87	89
82	85
80	78
75	73
65	67
50	45
20	40

a. Give the equation of the linear regression line for this set of data.

b. A new student comes to the class and earns a homework grade of 78. Based on the equation in part a, what grade would the teacher predict the student would receive on the subsequent test, to the nearest integer?

<u>Sample #10:</u> The points in the scatter plot at the right represent the ages of automobiles and their values. Based on this scatter plot, it would be reasonable to conclude:

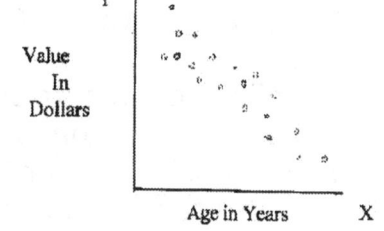

Value
In
Dollars

Y

Age in Years   X

(1) Age and value have a coefficient of correlation that is less than zero.
(2) Age and value have a coefficient of correlation that is equal to zero.
(3) Age and value have a coefficient of correlation that is between zero and 0.5.
(4) Age and value have a coefficient of correlation that is greater than 0.5.

Calculating Time:
Hours, Minutes, Seconds

Recall that we can enter angles or other measures in degrees, minutes, and seconds. While we will return to angle measurements later in the course, we can make use of the same methods to calculate time.

Because there are 60 seconds in a minute whether the minute is a measurement of time or a part of a degree, and there are 60 minutes in both an hour and a degree, we can enter hours, minutes, and seconds as degrees, minutes, and seconds.

Let's try it on this problem:
(Data source: The World Almanac and Book of Facts 2004)
In the 2002 New York City Marathon, the men's winner was Rodgers Rop of Kenya with a time of 2:08:07. The women's winner was Joyce Chepchumba, also of Kenya, with a time of 2:25:56. How much faster was the men's winner than the women's winner?

To find the difference in their times:
1. Start on the home screen with the number of hours for the women's time.
2. Press 2nd APPS to find the ANGLE menu and choose 1: °.
3. Follow the degree symbol with the number of minutes.
4. Go back to the ANGLE menu and choose 2:'.
5. Follow the minute symbol with the number of seconds.
6. Press ALPHA, + for the seconds symbol.
7. Subtract the men's time input in the same manner.
8. You should have a decimal answer.
9. To convert the decimal answer to hours, minutes, and seconds, return to the ANGLE menu and choose 4:>DMS.
10. The difference between Joyce's time and Rodgers' time is

_____ .

Practice:

1. The Ironman Triathlon World Championships - a 2.4-mile ocean swim, 112-mile bike ride and 26.2-mile run - are held annually at Kailua-Kona, Hawaii. The first winner, Gordon Haller in 1978, finished with a time of 11:46:58. In 2002, Timothy Deboom won with a time of 8:29:56. How much faster was Deboom than Haller? _____

2. Natasha Badmann of Switzerland was the women's Ironman Triathlon winner in 2002. Her time was 9:07:54. What was the difference between Natasha's time and Timothy's time?

_____

3. The last time the United States won the 4-Man Bobsled race in the Winter Olympics was in 1948 with a time of 5 minutes, 20.1 seconds. In 2002 Germany won with a time of 3 minutes, 7.51 seconds. How much faster (to the nearest tenth of a second) was this German team than the 1948 United States team? (Note: If there are no hours, 0° must be used to calculate the time.)

_____

4. In the 1928 Winter Olympics, Per Erik Hedlund of Sweden won the Men's 50 Kilometer Cross-Country Skiing event with a time of 4 hours, 52 minutes, and 3.0 seconds. In 2002 Mikhail Ivanov of Russia won this event with a time of 2 hours, 6 minutes, and 20.8 seconds. How much longer, to the nearest tenth of a second, did it take Hedlund to complete the race than Ivanov? _____

5. In the 2002 Winter Olympics Jochem Uytdehaage of the Netherlands won the 10,000 meter Men's Speed Skating Event setting an Olympic record with a time of 12 minutes, 58.92 seconds. The last winner of this event from the United States was Eric Heiden at the 1980 Winter Olympics at Lake Placid, NY. Heiden's time was 14 minutes, 28.13 seconds. How much longer than the new record, to the nearest hundredth of a second, did it take Heiden to complete the course?

_____

<u>Aug '01, #14:</u> A cellular telephone company has two plans. Plan A charges $11 a month and $0.21 per minute. Plan B charges $20 a month an $0.10 per minute. After how much time, to the nearest minute, will the cost of plan A be equal to the cost of plan B?

(1) 1 hr 22 min	(3) 81 hr 8 min
(2) 1 hr 36 min	(4) 81 hr 48 min

Quadratic Regressions

Quadratic regressions were covered in the first course, but here's a little refresher in case you've forgotten them:

To complete a quadratic regression, you must know at least _three_ points on the parabola.

1. Enter the x-coordinates in L_1.

2. Enter the y-coordinates in L_2.

3. Press STAT, go over to CALC, and choose 5:QuadReg.

4. Press ENTER. (Note: If your lists are entered in lists other than L_1 and L_2 or the coordinates are switched, you must follow QuadReg on the home screen with the list names with a comma between them.)

5. a, b, and c will be listed below the general form of a quadratic equation. Write the equation with these values appropriately substituted.

Try this one: Find the equation of the parabola that passes through the points (-1,-14), (0,-7), and (3,-10).

Let's try a new twist. In Math B we see lots of projectile questions. In each of these questions, "a" or the coefficient on x^2, represents the force of gravity acting on the object. You will notice that this number is –16 when the unit of measurement of the height is in feet, -4.9 when the unit of measurement is in meters.

"b" or the coefficient on x, represents the initial velocity of the object.

"c" is the height from which the object is initially dropped or thrown. (Quite often this is ground level, or 0).

If we can identify three points in the objects path by time and height where time is the independent (x) variable and height is the dependent (y) variable, then we can calculate the regression equation to find the object's initial velocity.

Practice:

1. Jake is a circus clown who performs a human cannonball act. He is fired from a cannon at ground level, reaches a maximum height of 30 feet after 2 seconds and lands in a ground level net 4 seconds after being "fired". What is Jake's velocity as he leaves the cannon?

2. A little boy is riding the Ferris wheel with his mother. Their seat is stopped at the top to let on new riders. He decides to start swinging his feet to make his mother nervous but accidentally throws his shoe. If the Ferris wheel is 264 feet high, the shoe reaches a maximum height 2 feet up and 3 feet over from the Ferris wheel and lands 5'10" from the point on the ground directly below the boy, what equation best represents the shoe's flight?

3. A football is kicked from the 30-yard line of one end of the football field and lands 4 seconds later on the 15-yard line at the other end of the field. If the football reaches a maximum height of 64 feet, what was the initial velocity of the football?

4. Cupid fires an arrow that reaches a maximum height of 100 feet and hits its target 5 seconds after being fired. What is the equation that represents the arrow's flight and what is its initial velocity?

Regression Practice

Use the data below to answer the questions that follow.

Month:	Average High:	Average Low:	Mean:	Average Sunrise:	Average Sunset:	Length of Day:
Jan	27	8	17	7:33	4:50	
Feb	30	10	20	7:01	5:32	
Mar	39	21	30	6:14	6:08	
April	52	32	42	6:19	7:46	
May	67	44	55	5:36	8:21	
June	75	52	63	5:19	8:46	
July	79	56	68	5:33	8:42	
August	77	54	66	6:06	8:06	
Sept	69	47	58	6:41	7:13	
Oct	56	37	47	7:16	6:18	
Nov	44	29	36	6:56	4:36	
Dec	32	15	24	7:30	4:24	

Data source: http://www.weather.com 10-22-02 Based on at least 30 years of records for the Castorland area (Watertown/Fort Drum).

1. Begin by finding the average length of day and fill in the last column of the chart.

2. Using the length of day as the independent variable, decide whether there is a good correlation between length of day and mean temperature for the month. Include the correlation coefficient in your response.

    ----

3. Again using the length of day as the independent variable, find whether there is a better correlation between the length of day and the average high temperature, the average low temperature, or the mean temperature. Include all three correlation coefficients in your response.

_____

_____

_____

****Do #4 and 5 at the same time to eliminate duplication of work.

4. Using quadratic regressions, find the equation of the curve of best fit for each of the following: (Use the number of the month as the independent variable.)

    a. Average High Temperature: _____

    b. Average Low Temperature: _____

    c. Mean Temperature: _____

    d. Sunrise: _____

    e. Sunset: _____

    f. Length of Day: _____

5. A quadratic regression does not have a correlation coefficient. For each of the above create a scatter plot on the graphing calculator and graph the equation found on the same screen. Decide visually whether it is a good fit or not, then record your decision and reasoning on the next page. You may include a sketch if it helps to explain whether it is a good fit or not.

a. Average High Temperature:

----

b. Average Low Temperature:

----

c. Mean Temperature:

   ----

d. Average Sunrise:

----

e. Average Sunset:

----

f. Average Length of Day:

Power Regressions

Power regressions will be executed exactly as linear and quadratic regressions are.

A power function is

_____

_____

To perform a power regression:
1. Enter the data in lists.
2. Turn diagnostics on in the Catalog.
3. Press STAT.
4. Go over to CALC.
5. Choose A:PwrReg.
6. On the home screen, if the independent and dependent variable data are entered in lists 1 and 2 respectively press enter. Otherwise follow PwrReg with the list names for the independent data and the dependent data separated by a comma.
7. A screen like the one at the right should appear.
8. "a" is the coefficient, "b" is the exponent, and "r" is the correlation coefficient.

```
PwrReg
 y=a*x^b
 a=213.8488542
 b=⁻.1023307424
 r²=.8992064364
 r=⁻.9482649611
```

What does "b" tell us about how the function will look if we graph it?

A quick exponent review may help:
   a. Exponents greater than 1

   _____

   b. Fractional exponents

   _____

   c. Negative exponents

   _____

In the example, b is a negative decimal so, as x increases, the function will _____

Practice:

Clara and Edna are planning an afternoon tea for their friends. Clara insists that a large teapot will keep the tea hot longer due to the larger volume. Edna argues that a small teapot will keep it just as warm and they can offer their friends a larger variety with less waste.

To settle the argument they perform an experiment to see who is right. While they are at it, they decide to check some other possibilities. The data below was collected using a tea thermometer (yes there is such a thing!) All three teapots and both cups are the same type of china. All three teapots are the same style and both teacups are the same style. The "covered" teapot is covered with a tea cozy between measurements. All actual teapot covers were returned to the pots between measurements.

Temperature in Degrees Fahrenheit

Time in Minutes	Small Teapot	Large Uncovered	Large Covered	Cup of Cocoa	Cup of Tea
0	195	190	193	180	184
3	184	179	185	162	169
6	182	171	178	155	159
9	176	167	174	146	149
12	170	162	168	138	140
15	165	157	167	132	132
18	155	155	160	127	129
21	155	151	159	121	123
24	154	148	154	117	119
27	150	143	150	112	112

Enter the data in lists 1-6.
The time will be the independent variable for each regression.

***Round all coefficients, exponents, and correlation coefficients to the nearest hundredth.

1. Use a power regression to find the equation of a curve that represents the rate of cooling for the small teapot. What happens on your first attempt?

    _____

2. The calculator will not allow a power regression where the independent variable is zero. To fix this, eliminate the zero time and the data for this time in each list.

3. Try the equation for the small teapot again.
    a. Equation: _____

    b. Correlation: _____

    c. Is it a good fit? _____

4. Large uncovered teapot:

    a. Equation: _____

    b. Correlation: _____

    c. Is it a good fit? _____

5. Large covered teapot:

    a. Equation: _____

    b. Correlation: _____

    c. Is it a good fit? _____

6. Cup of cocoa:

    a. Equation: _____

    b. Correlation: _____

    c. Is it a good fit? _____

7. Cup of tea:

    a. Equation: _____

    b. Correlation: _____

    c. Is it a good fit? _____

8. At what time will each teapot and cup reach room temperature according to your equations?

Small teapot: _____   Cup of Cocoa: _____

Large Uncovered teapot: _____   Cup of Tea: _____

Large Covered teapot: _____

Were the models good when extended over a longer time period? _____

Explain:

_____

_____

9. Edna and Clara decide on a large teapot with a tea cozy. Almost everyone will have ordinary black tea. They plan to allow 5 minutes for the tea to brew then 5 more minutes for the tea to cool. What temperature will the tea be at this time?

                                                    _____

10. One of their guests is an expectant mother who prefers decaffeinated tea. They decide to make a single cup for her. How much later should they begin to prepare her tea so that it will be the same temperature at the time they serve their guests?

                                                    _____

11. Two children will also be attending so Edna decides to prepare cups of hot cocoa for them. The cocoa will need to be cooler at serving time so she decides to prepare it 10 minutes before they begin to prepare the pot of tea. What will be the temperature of the cocoa when the tea is served?

_____

Variance

First some definitions:
Measure of Dispersion:

_____

_____

Deviation:

_____

_____

Now, *variance* is the _____ of the _____ of the

_____ from the _____ .

For those of you who enjoy grinding out unwieldy formulas, this is what the formula looks like:

$$v = \frac{\sum_{i=1}^{n}(x_i - \bar{x})^2}{n}$$

For those of you that prefer not to deal with Greek letters if you don't have to let's try it on the graphing calculator.

1. Enter the data in L_1.

2. Press 2nd, STAT, then go over to MATH.
3. Choose 8:variance(
4. Press 2nd, 1, then ENTER.

Because this is in an area we don't use often, it might be just as easy to find variance in the catalog.

Let's try it with some test scores:
Find the variance of the test scores in Mrs. Noftsier's graphing calculator class: 87, 99, 53, 54, 98, 100, 89, 96, 39, 67, 99, 91, 92, 93, 95, 96, 92, 47, 98, 96, 48.

_____

Practice:

1. Drew keeps track of how long he spends doing his math homework for a week: Monday 45 minutes, Tuesday 50 minutes, Wednesday 35 minutes, Thursday 40 minutes, and Friday 45 minutes. What is the variance in the length of time he spends on homework?

2. The following table gives the receipts for farm products for selected NYS counties for the year 2000. Find the mean, median, and variance for the data.

County	$$
Lewis	67,500
Franklin	48,499
Jefferson	82,451
Oneida	81,327
Oswego	41,259
St. Lawrence	96,521
Onondaga	76,610
Herkimer	49,919

Mean: _____   Median: _____ Variance: _____

3. The salaries in the table on the next page were listed in the February 15, 2004 edition of the *Watertown Daily Times*. Fill in the table below:

	Mean	Median	Variance
Supervisor			
Highway Superintendent			
Clerk			

Lewis County Town Salaries

Town	Supervisor	Highway Superintendent	Clerk
Croghan	$11,500	$35,000	$15,000
Denmark	$10,050	$37,000	$15,750
Diana	$8,500	$33,000	$13,500
Greig	$11,500	$32,300	$15,500
Harrisburg	$5,500	$24,500	$5,250
Lewis	$5,600	$30,000	$6,600
Leyden	$7,000	$28,900	$4,500
Lowville	$12,200	$40,900	$20,900
Lyonsdale	$7,500	$30,000	$2,300
Martinsburg	$8,000	$32,000	$11,000
Montague	$7,800	$11,000	$5,500
New Bremen	$10,800	$35,500	$15,000
Osceola	$3,600	$27,500	$5,000
Pinckney	$3,200	$24,500	$5,000
Turin	$7,500	$27,000	$9,500
Watson	$10,400	$36,200	$12,100
West Turin	$7,000	$31,000	$6,300

4. The table below gives the average snowfall in inches for the month of February. Find the variance.

Town	Snowfall Average
Carthage	24.3
Croghan	23.3
Fort Drum	24.3
Castorland	24.1
Lowville	24.3

Mean Absolute Deviation: Calculating With Lists

Definition: If $\bar{x}$ is the mean of a set of numbers denoted by x_i, then the mean absolute deviation, (or just the mean deviation) is:

$$\frac{\sum_{i=1}^{n}\left|x_i - \bar{x}\right|}{n} \qquad \text{Or} \qquad \frac{1}{n}\sum_{i=1}^{n}\left|x_i - \bar{x}\right|$$

Although it looks complicated, if we look closely at what is really going on it is a simple calculation.

1. **Find the mean** of the set of data.
2. **Find the difference between each individual piece of data and the mean** (how much it deviates from the mean). Because we are looking for a distance from the mean and distance is always positive, if there are any negative values make them positive (**absolute value**).
3. **Average these differences**.

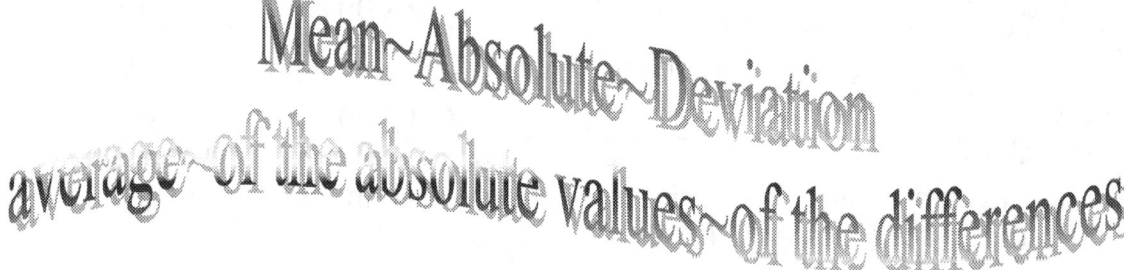

On paper this usually takes several minutes but lists can be manipulated to make our work *much easier*!

1. Enter the initial set of data in List 1.
2. Find the mean using 1-Variable Stats. You don't have to remember it, but the calculator has to have calculated it.
3. Return to the List Editor, move the cursor **up** and over so that L_2 is highlighted.   (Note that you need to move UP to use the bottom of your screen!)

4. Now enter the following expression: abs($L_1 - \bar{x}$).

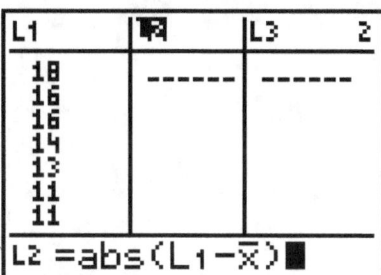

Need help with the keystrokes?
   a. MATH, over to NUM, choose 1:abs(
   b. 2nd, 1 will insert L_1.

   c. To subtract the mean, after the operation symbol, press **VARS, choose 5:Statistics..., then 2: $\bar{x}$.** This will use the calculated mean in your expression. (You could enter the value instead of the symbol if you prefer.)

5. Press ENTER. A new list of data should appear in L_2.

6. Find the mean of this set of data using 1-Variable Stats. This is your answer!

The table on the next page lists weather data for Croghan, NY. Use this table or find the data for a town or city near you to answer the questions below. You can find the data at http://www.city-data.com.

1. Find the mean absolute deviation for the monthly wind speed.

   _____

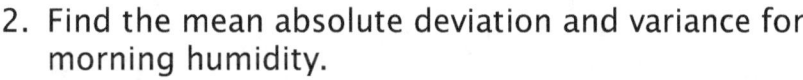

2. Find the mean absolute deviation and variance for morning humidity.

   _____   _____

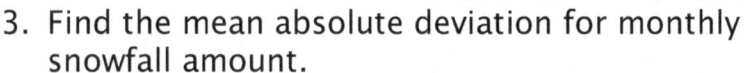

3. Find the mean absolute deviation for monthly snowfall amount.

   _____

4. Find the mean absolute deviation and variance for monthly high temperatures.

   _____   _____

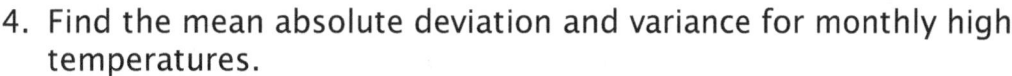

5. Find the mean absolute deviation and variance for monthly precipitation.

   _____   _____

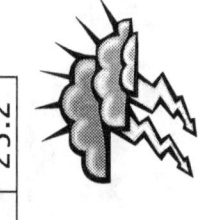

Average Weather and Climate for Croghan, NY

	Jan	Feb	Mar	Apr	May	Jun	Jul	Aug	Sep	Oct	Nov	Dec
Average Temp.	17	19.3	29	41.8	54.7	62.9	67.3	65.4	57.4	46.4	35.6	23.2
High Temp	26.6	29.2	38.3	51.7	66	73.9	78.4	76.3	67.8	56	43	31.4
Low Temp	7.2	9.5	19.7	31.8	43.3	51.7	56.2	54.5	46.9	36.7	28.1	14.9
Precipitation	4.1	2.8	3.3	3.4	3.5	3.6	3.9	3.9	4.5	3.8	4.6	4.1
Days with Precipitation	18	15	16	14	13	11	11	11	11	13	16	18
Wind Speed (mph)	10.4	10.4	10.6	10.3	9.0	8.3	7.9	7.5	8.1	8.6	10.0	10.2
Morning Humidity	77	78	78	75	76	79	81	86	87	84	81	81
Afternoon Humidity	68	63	59	53	54	56	56	58	61	61	66	70
Sunshine (%)	35	41	47	49	55	59	63	59	53	45	28	27
Days Clear of Clouds	3	3	5	6	6	7	8	7	7	6	2	2
Partly Cloudy Days	7	6	7	7	10	11	12	11	10	8	6	5
Cloudy Days	21	18	19	17	15	12	11	13	13	17	22	23
Snowfall (in)	28.0	23.3	17.2	3.6	0.1	0.0	0.0	0.0	0.0	0.5	9.3	25.2

Graphing Calculator for NYS Math B and Beyond

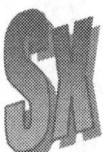 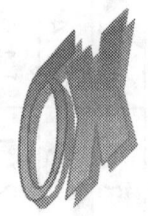

Standard Deviation

Data used for statistics can be found in two ways.

Everyone from the group can be included.
This would be the whole _____ .

Or

A few people can be chosen to represent the whole group.
This would be a _____ .

The best statistics include the whole group. The next best is

----

Standard Deviation is _____

The formula for standard deviation is:

$$\sigma = \sqrt{\frac{\sum_{i=1}^{n}(\bar{x} - x_i)^2}{n}}$$

Where $\bar{x}$ is the _____ ; x_i is the _____ ;
n is _____ .

Compare the formulas for standard deviation and variance.

What is the relationship between the variance and the standard deviation?

Example:

Five candy bars are compared for grams of total fat contained in a 64-gram bar:

Milk Chocolate: 21g
Crisped Rice: 20g
Almond: 23g
Peanut Butter: 19g
Mint Cookie: 20g

To calculate standard deviation on the TI-83+
1. Enter the data in a list.

2. Press | STAT |

3. Move over to CALC.

4. Chose 1:1-Var Stats.

5. a. If there is only one list press | ENTER | .

b. If there are two or more lists on your calculator give the list name after the prompt on the home screen before pressing | ENTER | .

6. a. If the data represented a _____, then σX is the appropriate standard deviation.

b. If the data represented a _____, then Sx is the appropriate standard deviation.

Find the standard deviation for grams of fat in the candy bars:

_____

How do I tell if it is a sample or a population?

A word of warning!! Only do standard deviation in 1-variable stats if it is a population!!!!! If you find standard deviation in the catalog or even at 2nd STAT, it will be for the sample only!!!

Practice:

1. Which of the following would indicate that a sample was being used?
    a. Mrs. Noftsier graded 10 tests and found that the standard deviation was 10.4.
    b. Mrs. Noftsier checked 10 of the 60 tests and found that the standard deviation was 10.4.

    ----

2. Santa has asked for a report on how the reindeer are eating. He receives the following report:

Donder	5lbs of grain
Blitzen	4.5 lbs of grain
Comet	5.25 lbs of grain
Dancer	4.8 lbs of grain

What is the standard deviation?

#3-8 Use the data in the table below to:

3. Find the mean, median, and mode.

---------- ---------- ----------

4. Find the variance.

----

5. Find the standard deviation.

----

6. Find the mean absolute deviation.

----

7. Find the equation of the line of best fit.

----

pH measurements of Precipitation on Whiteface Mountain:

1989	4.44
1990	4.36
1991	4.34
1992	4.46
1993	4.32
1994	4.42
1995	4.54
1996	4.60
1997	4.47
1998	4.48
1999	4.56
2000	4.52
2001	4.56

8. Find the correlation coefficient and give an explanation.

---------- --

Grouped Data:
Measures of Central Tendency

Sometimes data isn't given in an item-by-item list. If data is "grouped" we must treat it differently.

Measure of Central Tendency:

_____

_____

Frequency: _____

Suppose that we start with the data below:

Data	Frequency
75	2
80	5
85	8
90	6
95	3
100	1

We could rewrite this as one list with 75 appearing twice, 80 appearing 5 times, etc.

Yes, there is another way!

Enter the data in List 1 and the frequency in List 2.
First we will concentrate on measures of central tendency. (We will find the _____ and the _____ .)

1. Enter the data in L_1 and the frequency in L_2.

2. Return to the home screen.
3. Press 2nd, STAT, go over to Math and choose #3:mean(.
4. Enter the data list first.
5. Use a comma after the data list.
6. Enter the frequency list.

```
NAMES OPS MATH
1:min(
2:max(
3:mean(
4:median(
5:sum(
6:prod(
7↓stdDev(
```

7. Press ENTER.

The calculator will automatically convert the lists and find the mean of the entire set of data.

The mean of the set of data in the table is: _____ .

Similarly, we can find the median of the data. Use #4:median(and enter the data, a comma, then the frequency.

The median of the set of data in the table is: _____ .

Note: This method will also work with the 1-Variable Stats.

Practice:

1. The table below shows the scores of 40 students on an advanced placement mathematics examination. Find the mean and median.

Score	Number of Students
5	7
4	13
3	14
2	3
1	3

2. In the table of weather and climate for Croghan, NY, monthly averages were listed. To convert these to daily averages, we can list the average in L_1 and the number of days in the month in L_2 as our frequency list.

    a. Find the daily average high temperature. _____

    b. Find the daily average precipitation. _____

    c. Find the daily average afternoon humidity. _____

    d. Find the daily average snowfall. _____

Grouped Data: Measures of Dispersion

Just like the measures of central tendency, the measures of dispersion can be calculated using a list of frequencies and a list of data on the TI-83+/TI-84+.

The measures of dispersion we have looked at so far are _____ , _____ and _____ .

Some new notation:
When working with grouped data we might see the data in a table like the one below:

x_i	f_i
15	4
20	7
25	13
30	6

The first column (x_i) lists the data as it would appear in a complete list. The second column (f_i) is a frequency list. For example, this table tells us that the number 15 would appear 4 times if we had the complete list of data.

Find the variance and the standard deviation for the data in the table above. Remember that the standard deviation should be found in the 1-Variable Stats to be sure you have the correct one!

Standard deviation = _____ variance = _____

Practice:

1. The table at the right shows raw scores on an 80-question entrance examination. Find the standard deviation of these examination scores to the nearest tenth.

_____

x_i	f_i
40	5
50	4
60	6
70	3
80	2

2. Using the weather and climate data for Croghan, NY and the method described in the previous lesson,
   a. Find the standard deviation and variance for the daily average temperature. _____ _____
   b. Find the standard deviation and variance for the daily high temperature. _____ _____
   c. Find the standard deviation and variance for the daily low temperature. _____ _____
   d. Find the standard deviation and variance for the daily precipitation. _____ _____

Mixed Practice:

3. Find the mean absolute deviation for particulate matter in the cities listed in the table "Air Pollution in Selected Cities" (your list named PART). _____

4. Find the variance for sulfur dioxide in the selected cities. (SULF) _____

5. Find the standard deviation for nitrogen dioxide in the selected cities. (NITRO) _____

6. Find the mean absolute deviation and variance for each of the following lists:
   a. DEBT: _____ _____
   b. PCAP: _____ _____
   c. INT: _____ _____

7. Find the mean, mean absolute deviation, standard deviation, and variance for the total participants in wildlife associated recreation for the selected states. (Box-and-Whisker Plot lesson)
   a. Mean: _____
   b. Mean absolute deviation: _____
   c. Standard deviation: _____
   d. Variance: _____

Measures of Dispersion

June '02, #21: On a nationwide examination, the Adams School had a mean score of 875 and a standard deviation of 12. The Boswell School had a mean score of 855 and a standard deviation of 20. In which school was there greater consistency in the scores? Explain how you arrived at your answer.

June '02, #27: An electronics company produces a headphone set that can be adjusted to accommodate different-sized heads. Research into the distance between the top of people's heads and the top of their ears produced the following data, in inches:

4.5, 4.8, 6.2, 5.5, 5.6, 5.4, 5.8, 6.0, 5.8, 6.2, 4.6, 5.0, 5.4, 5.8

The company decides to design their headphones to accommodate three standard deviations from the mean. Find, to the nearest tenth, the mean, the standard deviation, and the range of distances that must be accommodated.

Aug '01, #11: On a trip, a student drove 40 miles per hour for 2 hours and then drove 30 miles per hour for 3 hours. What is the student's average rate of speed, in miles per hour, for the whole trip?

   (1) 34   (2) 35   (3) 36   (4) 37

Jan '03, #21: Two social studies classes took the same current events examination that was scored on the basis of 100 points. Mr. Wong's class had a median score of 78 and a range of 4 points, while Ms. Rizzo's class had a median score of 78 and a range of 22 points. Explain how these classes could have the same median score while having very different ranges.

<u>Jan '04, #6:</u> Jean's scores on five mathematics tests were 98, 97, 99, 98, and 96. Her scores on five English tests were 78, 84, 95, 72, and 79. Which statement is true about the standard deviations for the scores?

(1) The standard deviation for the English scores is greater than the standard deviation for the math scores.
(2) The standard deviation for the math scores is greater than the standard deviation for the English scores.
(3) The standard deviation for both sets of scores are equal.
(4) More information is needed to determine the relationship between the standard deviations.

Cumulative Normal Distribution Function

Mrs. Noftsier has 85 students in her mathematics class.
The scores on the Chapter 3 test are normally distributed and have a mean of 72.3 and a standard deviation of 8.9. How many students in the class can be expected to receive a score between 82 and 90?

Use your graphing calculator to find the area of the normal curve that lies between the lower abound of 82 and the upper bound of 90:

1. Press 2nd, VARS and choose 2:normalcdf(to select the normal

cumulative density function.

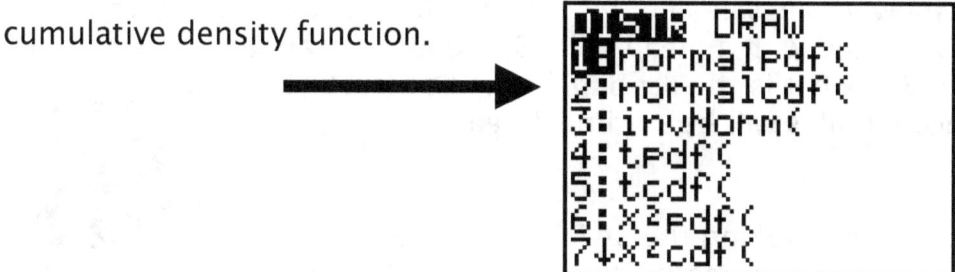

2. Enter the **lower bound**, the **upper bound**, the **mean**, the **standard deviation**, and the right parenthesis. Separate numerical values with commas:

3. Press ENTER to get .1145178018.

This value represents the percent of scores between 82 and 90. Since there are 85 students in the class, 85 *.1145≈10 students can be expected to receive scores between 82 and 90.

Practice:

1. On a Chemistry exam, the scores formed a normal distribution. If the mean of the scores is 80 and the standard deviation is 6, what percentage of the scores lies between 80 and 86? _____ If 55 students took the test, how many students scored between 80 and 86? _____

2. If syrup production per year in the United States is normally distributed with a mean of 4,660,000 liters and a standard deviation of 702,000, what is the probability that next years production will exceed 5,000,000 liters? _____ Over the next 20 years, in how many years would you expect the production to exceed 5,000,000 liters? _____

3. If snowmobile sales per year in the United States are normally distributed with a mean of 136,000 and a standard deviation of 30,000, what is the probability that at least 120,000 snowmobiles will be sold next year? _____

4. Of students taking the SAT I in New York State in 2003, the mean math score for students that reported that they use a calculator almost every day was 531 with a standard deviation of 108. Students that reported using a calculator only once or twice weekly had a mean math score of 484 with a standard deviation of 109. Assuming a normal distribution of scores, if a student uses a calculator almost every day, what is the probability that they will score at least a 600 on the Math portion of the SAT I? _____ What is the probability that a student that uses a calculator only once or twice a week will score at least a 600? _____

5. 23,167 students taking the SAT I in New York State in 2003 reported that their parents had a Bachelor's degree. These students had a mean Math score of 530 with a standard deviation of 105. Assuming a normal distribution, how many of these students scored between 500 and 600 on the Math test? _____

We have already investigated how to use the "normalcdf" to find the percent of data within a certain range. Now we will look at how to graph or draw the normal distribution curve.

The first choice on the distribution menu is normalpdf or _____

_____ _____ _____ . The difference between

this choice and the normalcdf is _____

To find "good" window for this type of graph, start with an Xmin and Xmax just below and above the data range and use Ymin=0 and Ymax=.08.

1. Go to Y=.
2. With your cursor on Y_1 press 2nd then VARS.
3. Choose 1: normalpdf.
4. After normalpdf(in Y_1 input x,mean,standard deviation).

5. If you do not enter a mean and standard deviation the calculator will use 0 as the mean and 1 as the standard deviation. (This will give you a copy of the basic normal curve.)

Practice:
1. Graph the normal curve for a set of data with a mean of 15 and a standard deviation of 2.5.

2. Graph the normal curve for a set of data with a mean of 4.02 and a standard deviation of 1.28.

3. Graph the normal curve for a set of data with a mean of 76 and a standard deviation of 9.6.

To draw the normal curve with a section shaded:
1. Press 2ⁿᵈ, VARS.
2. Go over to DRAW.
3. Choose 1: ShadeNorm(.
4. On the home screen input the low, high, mean and standard deviation after ShadeNorm(.
5. The area will be displayed with the picture of the shaded curve. This area represents a percent of data within the stated range or the probability that data will fall within that range.

Practice:

A. The weight in pounds per cubic foot of 96 common woods is normally distributed about a mean of 37.1 with a standard deviation of 8.6. What is the probability that someone with no knowledge of wood density randomly chooses a wood with a weight between 36 and 45 pounds per cubic foot. Sketch and shade the curve. _____

More Normal Curves

1. A certain bank is busiest during the Friday evening rush hours from 3:00 P.M. until 6:00 P.M.. During these hours the waiting time for drive-through customers is normally distributed with a mean of 6 minutes and a standard deviation of 1.5 minutes.

   a. What percent of drive-through customers will wait for 8 minutes or longer during the Friday evening rush hours?

   _____

   b. What is the probability that a customer will wait between 4 and 12 minutes during the Friday evening rush hours?

   _____

   c. What is the probability that a customer will wait 2 minutes or less during the Friday evening rush hours?

   _____

2. A company produces light bulbs having a life expectancy that is normally distributed with a mean of 950 hours and a standard deviation of 45 hours.

   a. What percent of the bulbs will last for 950 hours or more?

   _____

   b. What is the probability that a randomly chosen bulb will burn out in 905 hours?

   _____

   c. What is the probability that a randomly chosen bulb will last between 850 and 1050 hours?

   _____

3. According to a survey by the National Center for Health Statistics, the heights of adult women in the United States are normally distributed with a mean of 64 inches and a standard deviation of 2.7 inches.

   a. What is the probability that a woman selected at random is 57 inches or shorter?

   _____

b. What is the probability that a randomly selected woman has a height between 60 and 66 inches?

_____

c. What is the probability that a randomly selected woman has a height of at least 71 inches?

_____

4. The time that it takes for a particular fire department to arrive at a certain address on an emergency call is normally distributed with a mean of 12 minutes and a standard deviation of 2 minutes. What is the probability that the fire department takes longer than 12 minutes to arrive?

_____

5. A normal distribution has a mean of 36 and a standard deviation of 5. find the probability that a randomly selected *x*-value is in the given interval.
   a. Between 31 and 41   _____

   b. Between 21 and 36   _____

   c. Between 31 and 46   _____

   d. Less than 41   _____

   e. Greater than 26   _____

   f. Less than 51   _____

   g. Between 21 and 46   _____

   h. Greater than 36   _____

6. According to the International Snowmobile Manufacturers Association, worldwide snowmobile sales from 1992 to 2001 were normally distributed with a mean of 213,521 units and a standard deviation of 37986 units. If maximum production is 270,000 units, what is the probability that demand will exceed production?

_____

Inverse Normal
Inverse Normal

```
Area=.839995
low=55        up=65
```

The inverse normal probability density function will find the precise value at a given percent based on the mean and standard deviation.

1. Press 2ⁿᵈ VARS (DISTR).
2. Choose 3:invNorm(
3. Input the percent as a decimal followed by the mean then the standard deviation, with commas separating the values.
4. Press ENTER.

```
DISTR DRAW
1:normalpdf(
2:normalcdf(
3:invNorm(
4:tpdf(
5:tcdf(
6:X²pdf(
7↓X²cdf(
```

This will give the exact value for the given percent. Depending on the question, you may want values greater than or less than this value.

Examples:

1. The mean verbal score for juniors in New York State taking the PSAT in the fall of 2003 was 47.1 with a standard deviation of 10.6. Find the score represented by the following percentiles if the scores are normally distributed:

   a. 50%   _____

   b. 25%   _____

   c. 20%   _____

   d. 75%   _____

   e. 90%   _____

   f. 99%   _____

2. The mean math score for juniors in New York State taking the PSAT in the fall of 2003 was 48.2 with a standard deviation of 10.8. Find the range of scores achieved by less than 3% of the students taking the test.

   _____

Practice:

1. The national mean verbal score for juniors taking the PSAT in the fall of 2003 was 47.2 with a standard deviation of 10.8. Find the score that represents each of the following percentiles:

   a. 25%　　　_____

   b. 75%　　　_____

   c. 90%　　　_____

   d. 95%　　　_____

2. Males taking the PSAT in New York State had a mean math score of 49.8 with a standard deviation of 11.2. Females had a mean math score of 46.8 with a standard deviation of 10.2. Find the male and female scores at the following percentiles then find their difference:

   a. 25%　　_____ _____　　　_____

   b. 50%　　_____ _____　　　_____

   c. 75%　　_____ _____　　　_____

   d. 95%　　_____ _____　　　_____

In a study of 12 dairy farms the following results were found:

	Mean	Standard Deviation
Milk yield per cow (lbs/day)	62.9	21.3
Milk fat %	3.48	0.8
Milk protein %	3.39	0.4

Assume a normal distribution.

3. If John says his prize Holstein performs at the 95th percentile or better in each category, find the minimum amount of milk John's cow produces. What percent of fat and protein must this milk contain?

   _____ _____ _____

4. John's neighbor is known to be a bit of a cheapskate. As a result of the poorer quality feed he provides for his herd, his best cow is only at the 45th percentile. Find the yield for this cow and the percent of fat and protein her milk contains.

   _____ _____ _____

<u>June '02, #6:</u> On a standardized test, the distribution of scores is normal, the mean of the scores is 75, and the standard deviation is 5.8. If a student scored 83, the student's score ranks
    (1) below the 75th percentile
    (2) between the 75th percentile and the 84th percentile
    (3) between the 84th percentile and the 97th percentile
    (4) above the 97th percentile

<u>Sample #12:</u> The scores on a 100-point exam are normally distributed with a mean of 80 and a standard deviation of 6. A student's score places him between the 69th and 70th percentile. Which of the following best represents his score?

    (1) 66   (2) 81   (3) 84   (4) 86

<u>Sample #24:</u> A survey of the soda drinking habits of the population in a high school revealed the mean number of cans of soda consumed per person per week to be 20 with a standard deviation of 3.5. If a normal distribution is assumed, find an interval that contains the total number of cans per week approximately 95% of the population of this school will drink. Explain why you selected that interval.

<u>Aug '01, #29:</u> Twenty high school students took an examination and received the following scores:
  70, 60, 75, 68, 85, 86, 78, 72, 82, 88, 88, 73, 74, 79, 86, 82, 90, 92, 93, 73

Determine what percent of the students scored within one standard deviation of the mean. Do the results of the examination approximate a normal distribution? Justify your answer.

Jan '02, #26: A set of normally distributed student test scores has a mean of 80 and a standard deviation of 4. Determine the probability that a randomly selected score will be between 74 and 82.

June '01, #26: Professor Bartrich has 184 students in her mathematics class. The scores on the final examination are normally distributed and have a mean of 72.3 and a standard deviation of 8.9. How many students in the class can be expected to receive a score between 82 and 90?

June '03, #24: In a certain school district, the ages of all new teachers hired during the last 5 years are normally distributed. Within this curve, 95.4% of the ages, centered about the mean, are between 24.6 and 37.4 years. Find the mean age and the standard deviation of the data.

Aug '02, #2: In a New York City high school, a survey revealed the mean amount of cola consumed each week was 12 bottles and the standard deviation was 2.8 bottles. Assuming the survey represents a normal distribution, how many bottles of cola per week will approximately 68.2% of the students drink?

(1) 6.4 to 12   (3) 9.2 to 14.8
(2) 6.4 to 17.6   (4) 12 to 20.4

Aug '02, #22: The amount of time that a teenager plays video games in any given week is normally distributed. If a teenager plays video games an average of 15 hours per week, with a standard deviation of 3 hours, what is the probability of a teenager playing video games between 15 and 18 hours a week.

Jan '03, #8: The national mean for verbal scores on an exam was 428 and the standard deviation was 113. Approximately what percent of those taking this test had verbal scores between 315 and 541?

   (1) 68.2%   (2) 52.8%   (3) 38.2%   (4) 26.4%

Jan '03, #27: A shoe manufacturer collected data regarding men's shoe sizes and found that the distribution of sizes exactly fits the normal curve. If the mean shoe size is 11 and the standard deviation is 1.5, find:

   a. the probability that a man's shoe size is greater than or equal to 11

   b. the probability that a man's shoe size is greater than or equal to 12.5

   c. $\dfrac{P(size \geq 12.5)}{P(size \geq 8)}$

Aug '03, #17: The amount of ketchup dispensed from a machine at Hamburger Palace is normally distributed with a mean of 0.9 ounce and a standard deviation of 0.1 ounce. If the machine is used 500 times, approximately how many times will it be expected to dispense 1 or more ounces of ketchup?

   (1) 5   (2) 16   (3) 80   (4) 100

Jan '04, #11: Battery lifetime is normally distributed for large samples. The mean lifetime is 500 days and the standard deviation is 61 days. Approximately what percent of batteries have lifetimes longer than 561 days?

   (1) 16%   (2) 34%   (3) 68%   (4) 84%

At Most/At Least

When calculating "at most" and "at least" probabilities, we must not only consider the given probability, but also:

    a. all those smaller when it is "at most"

    b. all those larger when it is "at least"

Recall that $_nC_r$ finds the number of _____ of _____ things taken in groups of _____ . In probability it helps us to find the number of possible outcomes.

It is found by pressing MATH, going over to PRB, then choosing #3. Remember that the "n" needs to be entered on the homescreen before $_nC_r$ is entered then enter the "r" value.

The probability that an event will occur r times out of n trials can be calculated with this formula:

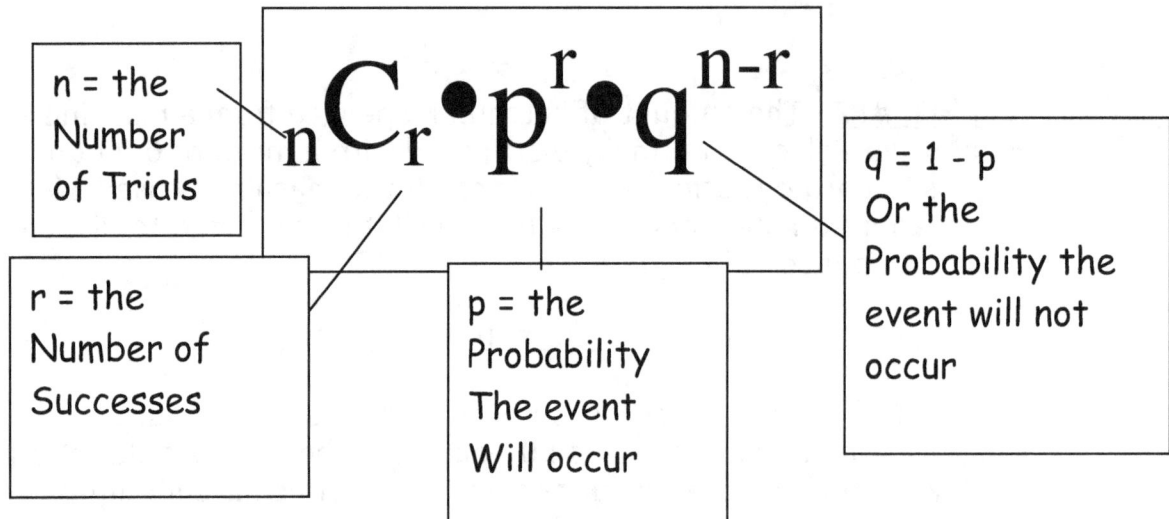

Let's try just plugging in some numbers first. Suppose n=5, r=3, and p=40%.

$$_nC_r(p^r)(q^{n-r})=\_\_C\_\_(\_\_\_)^{..}(\_\_\_)^{..}$$

$$=_____$$

Suppose a jar contains 3 red marbles, 6 blue marbles, and 2 green marbles. What is the probability of drawing a blue marble at least 3 out of 5 times?

Then you could draw a blue marble ____ times, _____times, or _____ times.

Then you must find $_nC_r p^r q^{n-r}$ for r=3, r=4, and r=5 and add these three probabilities for your final answer.

If r=3:

If r=4:

If r=5:

The sum= _____

What is the probability that you will draw a green marble at most 2 out of 6 times?

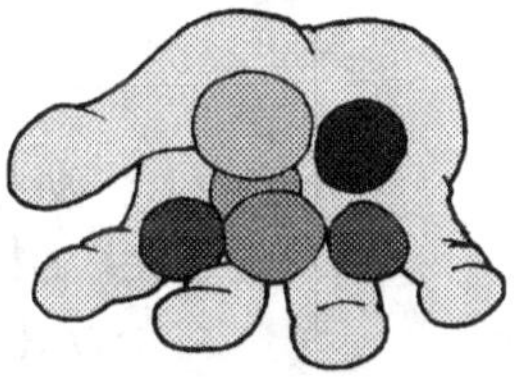

Practice:

1. Suppose your favorite baseball player is batting .365 at the time of a particular game. What is the probability that he will have a base hit at least 3 of 4 at bats in this game?

----

2. A relief pitcher strikes out 4 out of every 9 batters. The opposing team has the bases loaded with no outs when he enters the game. What is the probability that this pitcher will strike out the next three batters? (At least 3 of 3.)

----

3. Ann plays varsity basketball and is shooting 60% from the foul line. What is the probability that she will make at least 8 of 10 shots in her next game?

----

4. A coin has been tested and found to be weighted in such a way that it will land heads 2 of three tosses. What is the probability that it will land heads at most 5 of 10 tosses?

----

5. What is the probability that a family with 5 children will have at least 2 boys?

----

6. The weather forecast claims that there is a 50% chance of snow each of the next 5 days. What is the probability that it will snow at most 3 days?

Binomial Theorem

The binomial theorem allows us to use $_nC_r$ to save us work when expanding a binomial raised to a power.

Recall: A binomial is _____

_____

What is $(x + y)^2$?

What is $(x + y)^3$?

As the power on the binomial increases, it becomes more tedious to multiply.

The binomial theorem says:

$$(x+y)^n = {_nC_0}x^n y^0 + {_nC_1}x^{n-1}y^1 + {_nC_2}x^{n-2}y^2 + \ldots + {_nC_n}x^0 y^n$$

Try expanding $(x + y)^3$.

Now expand $(x + y)^4$.

How about $(x + y)^{10}$?

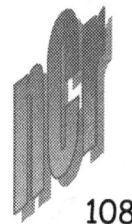

Practice:

1. Expand $(x + y)^7$.

2. Expand $(x + 1)^3$.

3. Expand $(2 + y)^5$.

4. Expand $(x + 7)^2$. Show ***two*** methods.

5. Expand $(x - 5)^4$.

6. What is the 4th term of the expansion of $(x+3)^8$?

7. What is the 3rd term of the expansion of $(x+2y)^6$?

8. What is the 2nd term of the expansion of $(7-y)^4$?

Using Lists to Calculate Probabilities

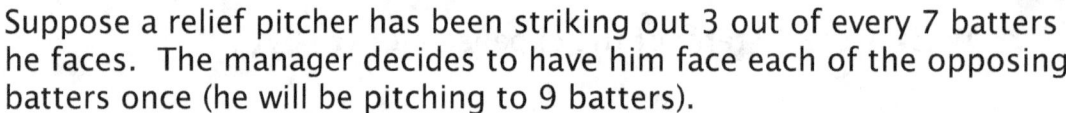

Lists can be used to calculate probabilities without using the formulas in the previous lessons.

Suppose a relief pitcher has been striking out 3 out of every 7 batters he faces. The manager decides to have him face each of the opposing batters once (he will be pitching to 9 batters).

1. Begin by entering the numbers 0 to 9 in list 1. This is only a reference list.

2. In list 2,
   a. Use the up arrow to highlight the name of the list.
   b. Press 2nd VARS.
   c. Choose 0:binompdf(
   d. Enter the number of trials, in this example use 9; followed by a comma and the probability that the event will occur, 3/7.
   e. This list is the *"EXACTLY"* list.

L1	🔳	L3	2
0	------	------	
1			
2			
3			
4			
5			
6			

L2 =...mpdf(9,3/7)▪

3. In list 3,
   a. Use the up arrow to highlight the name of the list.
   b. Press 2nd VARS.
   c. Choose A:binomcdf(
   d. Enter the number of trials, a comma, and the probability, just as in #2.
   e. This list is the *"AT MOST"* list.

L1	L2	🔳	3
0	.0065	------	
1	.04385		
2	.13155		
3	.23021		
4	.25898		
5	.19424		
6	.09712		

L3 =binomcdf(▪

4. In list 4,
   a. Use the up arrow to highlight the name of the list.
   b. Enter 1-L_3 as a formula.
   c. This is the *"AT LEAST"* list.

L2	L3	🔳	4
.0065	.0065	------	
.04385	.05035		
.13155	.18189		
.23021	.4121		
.25898	.67109		
.19424	.86532		
.09712	.96244		

L4 =1-L3

Fill in the table on the next page using the list entries and answer the questions that follow.

List 1	List 2	List 3	List 4
Number of Batters to Strike Out	Exactly	At Most	At Least

1. What is the probability that he will strike out exactly 6 of the 9 batters?
2. What is the probability that he will strike out at least 2 batters?
3. What is the probability that he will strike out at least 7 batters?
4. What is the probability that he will strike out at most 4 batters?
5. What is the probability that he will strike out at most 1 batter?

If a fractional response is more appropriate
1. Use the same "formula" as in the list. (For AT LEAST, use 1-binomcdf.)
2. Before pressing ENTER, press MATH, choose 1:>Frac.
3. Now press ENTER.   The list will appear in set notation with fractional entries.

Example:

If a fair coin is flipped 10 times, what is the probability that the coin will land on heads at least 7 times?

Use the right arrow to scroll to the correct place in the list. Be sure to begin counting with ZERO!

```
1-binomcdf(10,.5
)▶Frac
```

Note that some entries may remain in decimal form if the calculator is not able to convert them to an appropriate fraction.

Practice:

1. If a coin has been weighted so that it will land heads 4 times out of 5, find the probability that it will land tails at least 3 times out of 8 flips.

   _____

2. The weather report is calling for a 25% chance of rain each of the next 4 days.
   a. What is the probability that it will rain all 4 days?

   _____

   b. What is the probability that it will not rain any of the 4 days?

   _____

   c. What is the probability that it rains at least 2 of the 4 days?

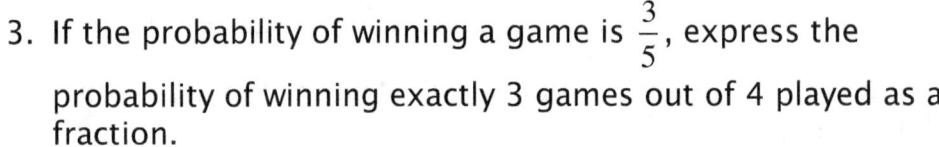

   _____

3. If the probability of winning a game is $\frac{3}{5}$, express the probability of winning exactly 3 games out of 4 played as a fraction.

   _____

4. In a family of six children,
   a. What is the probability that exactly one child is a girl?

   _____

   b. What is the probability that at least 2 of the children are girls?

   _____

5. Express the probability of rolling exactly three 4's in five rolls of a fair die as a fraction.

   _____

6. A basketball player is shooting 68% from the foul line. He is fouled on the last play of the game and his team is behind by one point. What is the probability that he will tie the game?
   What is the probability that he will score the winning point?

   _____

Using Lists with the Binomial Theorem:

Although expanding a binomial takes a little more work, we can still find the coefficients using the lists.

1. Make list 1 the numbers zero through the power the binomial is to be raised to.

2. In list 2 use $_nC_r$ using the power the binomial is to be raised to as the n and L_1 as the r.

This will give a quick list of the basic coefficients.

Practice:

1. Expand $(x+y)^7$.

2. Find the 5th term in the expansion of $(x-2)^8$.

   **To find the coefficient,

   a. Enter the numbers 0 through 8 in list 1

   b. Enter $_8C_{L1}$ as the formula in list 2.

   c. Enter $L_2*(-2)^\wedge L_1$ as the formula in list 3.

3. Expand $(x+5)^6$.

4. Find the 4th term of the expansion of $(x+y)^{10}$.

5. What is the third term in the expansion of $(x+1)^5$?

6. Expand $(2x-9)^9$.

<u>Jan '02, #29:</u> Team A and team B are playing in a league. They will play each other five times. If the probability that team A wins a game is 1/3, what is the probability that team A will win at least three of the five games?

<u>Aug '01, #28:</u> As shown in the accompanying diagram, a circular target with a radius of 9 inches has a bull's-eye that has a radius of 3 inches. If five arrows randomly hit the target, what is the probability that at least four hit the bull's-eye?

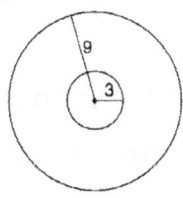

<u>June '01, #22:</u> At a certain intersection, the light for eastbound traffic is red for 15 seconds, yellow for 5 seconds, and green for 30 seconds. Find, to the nearest tenth, the probability that out of the next eight eastbound cars that arrive randomly at the light, exactly three will be stopped by a red light.

<u>June '02, #23:</u> After studying a couple's family history, a doctor determines that the probability of any child born to this couple having a gene for disease X is 1 out of 4. If the couple has three children, what is the probability the exactly two of the children have the gene for disease X?

Sample #18: A fair coin is tossed 5 times. What is the probability that it lands tails up exactly 3 times?

(1) $\left(\dfrac{1}{2}\right)^3$ (2) $\dfrac{3}{5}$ (3) $10\left(\dfrac{1}{2}\right)^5$ (4) $10\left(\dfrac{1}{2}\right)^3$

Sample #22: Jim can drive a golf ball over 220 yards 40% of the time. He regularly plays on a golf course where drives of that distance are needed on 12 holes. Determine the probability that exactly 7 of his drives will be over 220 yards.

June '03, #31: On any given day, the probability that the entire Watson family eats dinner together is $\dfrac{2}{5}$. Find the probability that, during any 7-day period, the Watson's eat dinner together at least six times.

Aug '02, #1: Which fraction represents the probability of obtaining exactly eight heads in ten tosses of a fair coin?

(1) $\dfrac{45}{1024}$ (2) $\dfrac{64}{1024}$ (3) $\dfrac{90}{1024}$ (4) $\dfrac{180}{1024}$

Aug '02, #8: What is the last term in the expansion of $(x+2y)^5$?

(1) y^5 (2) $2y^5$ (3) $10y^5$ (4) $32y^5$

<u>Jan '03, #2:</u> The probability that Kyla will score above a 90 on a mathematics test is $\frac{4}{5}$. What is the probability that she will score above a 90 on three of the four tests this quarter?

(1) ${}_4C_3\left(\frac{4}{5}\right)^3\left(\frac{1}{5}\right)^1$ (2) ${}_4C_3\left(\frac{4}{5}\right)^1\left(\frac{1}{5}\right)^3$ (3) $\frac{3}{4}\left(\frac{4}{5}\right)^3\left(\frac{1}{5}\right)^1$ (4) $\frac{3}{4}\left(\frac{4}{5}\right)^1\left(\frac{1}{5}\right)^3$

<u>Aug '03, #34:</u> When Joe bowls he can get a strike (knock down all the pins) 60% of the time. How many times more likely is it for Joe to bowl at least three strikes out of four times as it is for him to bowl zero strikes out of four tries? Round your answer to the nearest whole number.

<u>Jan '04, #28:</u> A board game has a spinner on a circle that has five equal sectors, numbered 1, 2, 3, 4, and 5, respectively. If a player has four spins, find the probability that the player spins an even number no more than two times on those four spins.

Complex Numbers

The TI-83+/TI-84+ will handle some complex functions in Real Number mode, but it is best to begin by changing to Complex Mode.

Press [Mode]

Using the arrow keys, move the cursor so that $a+bi$ is highlighted.

Press [ENTER]

Quit.

Enter $\sqrt{-25} + \sqrt{-16}$ in the calculator.

What answer does the calculator give?

> Note that the calculator automatically places beginning parentheses after the square root symbol. Be sure to *enter the closing parentheses* in the appropriate places!!

Try changing back to Real mode and entering the same expression. What happens?

_____

_____

**Change back to Complex mode for the remainder of the exercise

Press [MATH]

Move the cursor to CPX.

Your screen should now look like this:

```
MATH NUM CPX PRB
1▮conj(
2:real(
3:imag(
4:angle(
5:abs(
6:▶Rect
7:▶Polar
```

1. The calculator will find the Complex Conjugate of the complex number you place in the parentheses.

2. The calculator will tell you the Real part of the complex number entered in the parentheses.

3. Find the Imaginary part of the complex number entered.
4. Find the Absolute Value of a complex number.

The conjugate of a complex number is _____

_____

_____

The absolute value of a complex number is _____

_____

_____

On the home screen addition, subtraction, multiplication, and division of complex numbers can be done easily, but BE SURE TO PLACE PARENTHESES AROUND EACH COMPLEX NUMBER!!

If you haven't found it yet, i is found on the key with the decimal point.

Ex: Find the sum of $2-4i$ and $3+9i$.

Enter this in the calculator as: $(2-4i)+(3+9i)$

You should get an answer in complex form: _____

(Note: If you leave out the parentheses in addition you will get the same result, but for every other operation your result will be incorrect so it is better to be in the habit of grouping all the time.)

Powers of i :

The TI-83+/TI-84+ will give the powers of i, but will sometimes need to be adjusted.

This is a "quirk" of the calculator and does not need to keep us from using it for this purpose.

First note that when "E" appears on the calculator screen it means that the calculator has found it necessary to express an answer in scientific notation.

Try entering i^9

(Use ∧ for exponents other than 2.)

What does your screen say? _____

The calculator feels that we need to add 1E-13 to i.

But what is 1E-13? In usual scientific notation this would read 1×10^{-13}.

What is this number in standard form?

_____

Should we worry about this amount added to the value of i?

_____

_____

Fill in the table on the next page. Work with a partner, if possible, to save time.

Power of i :	Value using Remainder Method:	Value on TI-83+/TI-84+:
0		
1		
2		
3		
4		
5		
6		
7		
8		
9		
10		
11		
12		
13		
14		
15		
16		
17		
18		
19		
20		

What do you notice about the calculator value as compared to the actual value?

_____

_____

Practice:

1. $\sqrt{-16} + \sqrt{-169} =$ _____   2. $\sqrt{-1} + \sqrt{-225} =$ _____

3. $\sqrt{-9} + \sqrt{-400} =$ _____   4. $\sqrt{-1/16} - \sqrt{-9/25} =$ _____

Change #4 to a fraction using the calculator: _____

5. Find the sum of $4+6i$ and $2-8i$. _____

6. Find the sum of $3i$ and $5-7i$. _____

7. Find the difference between $1-5i$ and $3+7i$. _____

8. Find the difference between $2+10i$ and $9-i$. _____

9. Find the product of $5+i$ and $3-2i$. _____

10. Find the product of $3i$ and $2-8i$. _____

11. Find the complex conjugate of $9-8i$. _____

12. Find the complex conjugate of $3+4i$. _____

13. Find the complex conjugate of $i-3$. _____

14. Find the complex conjugate of $-i+2$. _____

15. Find the absolute value of $10+7i$. _____

16. Find the absolute value of $1-12i$. _____

17. Simplify i^{53}. _____

18. Simplify i^{144}. _____

19. Simplify i^{303}. _____

20. Simplify i^{1002}. _____

Complex Numbers

<u>June '02, #15:</u> What is the sum of $\sqrt{-2}$ and $\sqrt{-18}$?

    (1) $5i\sqrt{2}$ (2) $4i\sqrt{2}$ (3) $2i\sqrt{5}$ (4) $6i$

<u>Sample #28:</u> Solve the equation $x^2 = 6x - 12$ and express the roots in simplest $a + bi$ form.

<u>Aug '01, #22:</u> Show that the product of $a + bi$ and its conjugate is a real number.

<u>Jan '02, #19:</u> The expression $(-1 + i)^3$ is equivalent to

    (1) $-3i$ (2) $-2 - 2i$ (3) $-1 - i$ (4) $2 + 2i$

<u>Jan '02, #22:</u> Solve for x in simplest $a + bi$ form: $x^2 + 8x + 25 = 0$

June '01, #8: Fractal geometry uses the complex number plane to draw diagrams, such as the one shown in the accompanying graph.

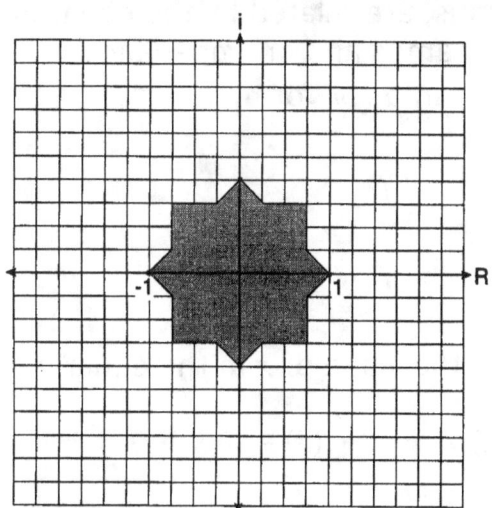

Which number is not included in the shaded area?
(1) $-0.5i$ (3) -0.9
(2) $-0.5-0.5i$ (4) $-0.9-0.9i$

June '01, #11: Melissa and Joe are playing a game with complex numbers. If Melissa has a score of $5-4i$ and Joe has a score of $3+2i$, what is their total score?

(1) $8+6i$ (2) $8+2i$ (3) $8-6i$ (4) $8-2i$

June '03, #4: The relationship between voltage, E, current, I, and resistance, Z, is given by the equation $E=IZ$. If a circuit has a current $I=3+2i$ and a resistance $Z=2-i$, what is the voltage of this circuit?

(1) $8+i$ (2) $8+7i$ (3) $4+i$ (4) $4-i$

June '03, #15: What is the value of $i^{99}-i^3$?

(1) 1   (2) i^{96} (3) $-i$ (4) 0

Aug '02, #15: Expressed in simplest form, $i^{16}+i^6-2i^5+i^{13}$ is equivalent to

(1) 1   (2) –1   (3) i (4) $-i$

<u>Jan '03, #25:</u> In an electrical circuit, the voltage, E, in volts, the current, I, in amps, and the opposition to the flow of current, called impedance, Z, in ohms, are related by the equation $E = IZ$. A circuit has a current of $(3+i)$ amps and an impedance of $(-2+i)$ ohms. Determine the voltage in $a+bi$ form.

<u>Aug '03, #2:</u> What is the value of x in the equation $\sqrt{5-2x} = 3i$?

    (1) 1   (2) 7   (3) –2   (4) 4

<u>Aug '03, #14:</u> What is the product of $5+\sqrt{-36}$ and $1-\sqrt{-49}$, expressed in simplest $a+bi$ form?

    (1) $-37+41i$ (2) $5+71i$ (3) $47+41i$ (4) $47-29i$

<u>Aug '03, #28:</u> Express, in simplest $a+bi$ form, the roots of the equation $x^2 +5 = 4x$.

Matrices

A matrix is a method for writing a system of equations in "shorthand".

The dimensions of a matrix are identified by how many rows high and how many columns wide it is.

The system $\begin{array}{c} -3x + 5y = 13 \\ 2x + 3y = 7 \end{array}$ can be written in a 2 x 3 matrix like the one below.

$$\begin{bmatrix} -3 & 5 & 13 \\ 2 & 3 & 7 \end{bmatrix}$$

Entering a Matrix:

To enter a matrix on the TI-83+/TI-84+:

1. Press 2nd ,x^{-1} (MATRIX)

2. Go over to EDIT.

3. Choose a matrix from A-J to edit.

4. Choose the number of rows then the number of columns (rows are horizontal, columns are vertical).

5. The correct number of spaces will be created in the matrix format. The default value for each entry is zero. Enter the values. Note that pressing ENTER will automatically take you to the next location in the matrix. The current location for the cursor is identified at the bottom of the screen by row,column=current value.

6. Return to the homescreen.

Discovering properties of matrices:

We can enter up to 10 matrices at a time so we will enter the matrices below and attempt operations on the pairs in the table. Look for patterns in error messages and make note of the type of error that is occurring.

$$A \begin{bmatrix} 1 & 4 \\ -2 & 0 \\ 7 & -3 \end{bmatrix} \quad B \begin{bmatrix} 4 & 2 & 0 \\ 3 & 1 & 5 \end{bmatrix} \quad C \begin{bmatrix} 2 & 5 \\ 4 & -3 \end{bmatrix} \quad D \begin{bmatrix} 1 & 2 & 7 \\ 0 & -4 & -8 \\ 3 & -1 & 4 \end{bmatrix} \quad E \begin{bmatrix} 1 & 0 \\ 0 & 1 \end{bmatrix}$$

$$F \begin{bmatrix} 1 & -4 & 7 \\ -9 & 11 & 0 \\ 3 & 10 & -1 \end{bmatrix} \quad G \begin{bmatrix} 0 & -1 \\ 1 & 0 \end{bmatrix} \quad H \begin{bmatrix} 3 & 10 \\ -7 & 2 \\ 0 & 6 \end{bmatrix} \quad I \begin{bmatrix} 2 & 7 \\ 3 & -1 \end{bmatrix} \quad J \begin{bmatrix} -2 & 2 & 0 \\ 5 & 6 & -1 \end{bmatrix}$$

To Perform Basic Operations on Matrices (+, -, x):
1. From the home screen, press 2nd ,x⁻¹(MATRIX).
2. Choose the first matrix from the list of NAMES.
3. Choose the operation.
4. Return to the MATRIX NAME menu if a second matrix is required.

To Perform Matrix Operations on Matrices:
1. From the home screen, press 2nd ,x⁻¹(MATRIX).
2. Move over to MATH.
3. Choose the necessary operation.
4. Return to the MATRIX menu.
5. Choose the matrix NAME for the matrix to be operated upon.

Note: An exponent of –1 means to _____ and is found using the x⁻¹ key.

Matrices	Operation	Result	Matrices	Operation	Result
A,B	+		A,H	+	
A,B	*		A,J	*	
A	x^{-1}		A,A	+	
A,C	+		A,A^{-1}	+	
A,C	*		A,A^{-1}	*	
A,D	+		C,C^{-1}	+	
A,D	*		C,C^{-1}	*	
B,D	+		D,F	+	
B,D	*		D,F	*	
B	x^{-1}		D	x^{-1}	
C,E	+		H,J	+	
C,E	*		H,J	*	
C	x^{-1}		I,E	+	
C,G	*		I,E	*	
E,G	*		E,E	*	
G,I	*		E	x^{-1}	
B,J	+		E,E^{-1}	*	
B,J	*		D,H	+	
J	x^{-1}		D,H	*	

Based on the results in the table, find the rules for adding, multiplying, and matrix inverse.

Addition:

_____

_____

Multiplication:

_____

_____

Inverse:

_____

_____

Practice:

1. What are the dimensions of the following matrices?

a. $\begin{bmatrix} 3 & -1 & 0 \end{bmatrix}$ b. $\begin{bmatrix} 1 & -5 \\ 6 & 10 \\ 3 & 7 \\ 2 & 0 \end{bmatrix}$ c. $\begin{bmatrix} 3 & 1 & -4 & 0 & 5 & 8 \\ 11 & 2 & 1 & -1 & 7 & 6 \\ 1 & 10 & 4 & 8 & 0 & -2 \end{bmatrix}$

_____   _____   _____

2. Add: $\begin{bmatrix} 2 & -3 & 4 \\ -1 & 0 & -3 \end{bmatrix} + \begin{bmatrix} -5 & 0 & -1 \\ 3 & 9 & 6 \end{bmatrix}$ _____

3. Multiply: $\begin{bmatrix} -1 & 7 & 3 \\ 0 & 7 & -2 \end{bmatrix} \times \begin{bmatrix} 3 & -1 \\ -2 & 4 \\ 1 & 5 \end{bmatrix}$ _____

4. Recall that an identity leaves the object it operates on unchanged.

   a. What is the identity matrix with respect to the operation <u>addition</u> for the matrix $\begin{bmatrix} a & b & c \\ d & e & f \\ g & h & i \end{bmatrix}$?   _____

   b. What is the identity matrix with respect to the operation <u>multiplication</u> for the matrix $\begin{bmatrix} 2 & -1 \\ -4 & 3 \end{bmatrix}$?

   _____

5. Find the inverse of the matrix $\begin{bmatrix} 2 & 3 & 1 \\ 1 & 2 & 3 \\ 3 & 1 & 2 \end{bmatrix}$. _____

***Use the following matrices for the remaining questions:

$A = \begin{bmatrix} 2 & 4 \\ 3 & -2 \end{bmatrix}$ $B = \begin{bmatrix} -2 & 5 \\ 4 & -3 \end{bmatrix}$ $C = \begin{bmatrix} -3 & 7 \\ 6 & -1 \end{bmatrix}$

6. Find $A \cdot A$.

7. Find $A \cdot A^{-1}$.   _____

8. Find $A + B$.   _____

9. Find $B + A$.   _____

10.   Find $A \cdot C$.   _____

11.   Find $C \cdot A$.   _____

12.   Find $A * (B + C)$.   _____

13.   Find $AB + AC$.   _____

14.   Do the associative, commutative, and distributive properties hold for matrix multiplication and addition? _____ Explain:

_____

_____

_____

Solving Systems With Matrices

As noted before, systems of equations can be entered in the calculator as matrices. We can then have the calculator convert the matrix to

_____ _____ _____ _____ . We will then have the solution if we know how to identify it.

In general, if the solution to the system $\begin{array}{c} ax+by=c \\ dx+ey=f \end{array}$ is (p,q), then the reduced row echelon form of the matrix $\begin{bmatrix} a & b & c \\ d & e & f \end{bmatrix}$ is $\begin{bmatrix} 1 & 0 & p \\ 0 & 1 & q \end{bmatrix}$.

Given two equations in two variables:

1. If the equations are not in the form $\begin{array}{c} ax+by=c \\ dx+ey=f \end{array}$, rearrange the terms so that they are in this form.

2. Enter the coefficients and constants in a 2 X 3 matrix.
   a. Press 2nd, x^{-1}.
   b. Choose EDIT.
   c. Choose 1:[A].
   d. Set the dimensions of the matrix as 2 X 3.
   e. If the numbers are entered in the following order pressing ENTER after each number will automatically take the cursor to the next position:
      i. Coefficient on x in the first equation.
      ii. Coefficient on y in the first equation.
      iii. Constant term in the first equation.
      iv. Coefficient on x in the second equation.
      v. Coefficient on y in the second equation.
      vi. Constant term in the second equation.

3. Return to the home screen.
4. Press 2nd, x^{-1}.
5. Choose MATH.
6. Choose B:rref(
7. Return to the MATRIX menu,2nd, x^{-1}.
8. Under NAMES, choose 1:[A].
9. Press ENTER.
10.   The solution to the system will be the last column of the matrix. The x value will be on top; the y value on the bottom.
11.   If these are decimals, round to the appropriate place value or press MATH. Choose 1:>Frac, ENTER.

Examples:

a. Find the solution to the following system:

$$3x - 7y = 12$$
$$-x + 9y = 17$$

----

b. Find the solution to the following system:

$$2x = -4y + 18$$
$$3x - 12y = 15$$

----

c. Write a system of equations for the problem below and solve.

Jenn's class is selling candles to raise money for the upcoming Prom. She collects the money when she delivers the candles but she forgets to bring the list telling how much each type of candle is and how much each customer owes. The first two customers remember what they owe. The first pays Jenn $46.45 for 2 small candles and 3 large candles. The second pays Jenn $49.50 for 4 small candles and 2 large candles. Jenn now knows how much to charge for each candle. How much should she collect for each small candle and each large candle from her remaining customers?

------------   ------------

Practice:

Convert solutions to fractions except in $ problems.

1. Solve: $3x - 5y = 12$
   $-4x + 11y = 17$

   _____

2. Solve: $12.5x + 18y = 25.3$
   $11.3x + 15y = 21.7$

   _____

3. Solve: $.85x - .7y = 10.9$
   $.15x + .9y = 9.4$

   _____

4. Steak and hamburg have been ordered for a family reunion cookout. It arrives in 2 boxes. One contains 8 packages of hamburg and 5 packages of steak and is marked 13.5 pounds. The second contains 6 packages of hamburg and 9 packages of steak and is marked 18 pounds. If each package of hamburg and each package of steak is the same weight, how much do the individual packages weigh?

   _____

5. A family went together to an amusement park but separated into two groups before lunch. When they got back together the parents compared "notes" and found that each group had decided on pizza and soda for lunch. One group spent $11.90 on 4 slices of pizza and 3 sodas. The other group spent $14.10 on 3 slices of pizza and 5 sodas. Dad is sure they ate at the same concession stand at different times. Mom is sure they couldn't have eaten at the same place. Who is right? Justify your conclusion.

   _____   _____

   _____

6. The owner of a florist shop had a family emergency and asked a friend, Andrea, to watch the shop for him. In his haste, he neglected to leave prices for his carnations and roses. Andrea found receipts showing that 15 roses and 10 carnations had been sold for $31.95 and 12 roses and 20 carnations had been sold for $35.60. A customer comes in and asks for 10 carnations and 6 roses. How much should Andrea charge for this order?

   _____

Solving Systems

<u>Aug '01, #14:</u> A cellular telephone company has two plans. Plan A charges $11 a month and $0.21 per minute. Plan B charges $20 a month and $0.10 per minute. After how much time, to the nearest minute, will the cost of plan A be equal to the cost of plan B?

(1) 1 hr 22 min (3) 81 hr 8 min
(2) 1 hr 36 min (4) 81 hr 48 min

<u>Jan '02, #28:</u> At the local video rental store, Jose rents two movies and three games for a total of $15.50. At the same time, Meg rents three movies and one game for a total of $12.05. How much money is needed to rent a combination of one game and one movie?

<u>June '02 #26:</u> Island Rent-a-Car charges a car rental fee of $40 plus $5 per hour or fraction of an hour. Wayne's Wheels charges a car rental fee of $25 plus $7.50 per hour or fraction of an hour. Under what conditions does it cost less to rent from Island Rent-a-Car?

<u>June '01 #23:</u> The cost of a long-distance telephone call is determined by a flat fee for the first 5 minutes and a fixed amount for each additional minute. If a 15-minute telephone call costs $3.25 and a 23-minute call costs $5.17, find the cost of a 30-minute call.

June '03, #28: The price of stock, $A(x)$, over a 12-month period decreased and then increased according to the equation $A(x) = 0.75x - 6x + 20$, where x equals the number of months. The price of another stock, $B(x)$, increased according to the equation $B(x) = 2.75x + 1.50$ over the same 12-month period. Graph and label both equations on the accompanying grid. State all prices, to the nearest dollar, when both stock prices are the same.

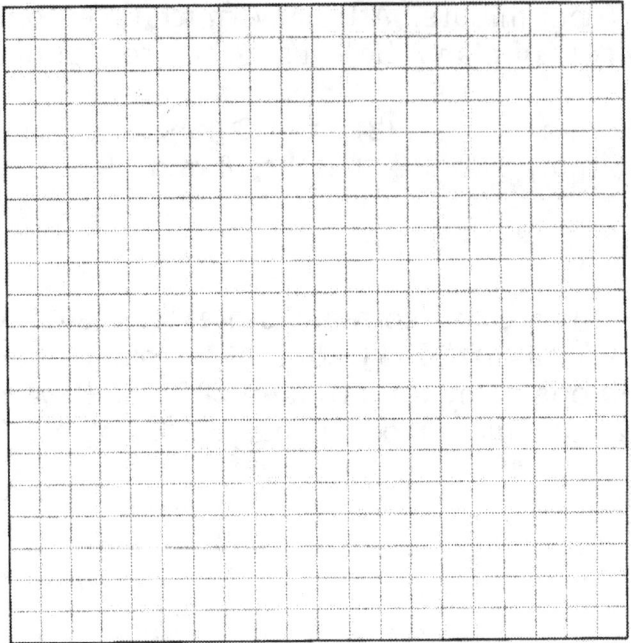

Aug '02, #27: The cost (C) of selling x calculators in a store is modeled by the equation $C = \dfrac{3,200,000}{x} + 60,000$. The store profit (P) for these sales is modeled by the equation $P = 500x$. What is the minimum number of calculators that have to be sold for profit to be greater than cost?

Aug '03, #32: A company calculates its profit by finding the difference between revenue and cost. The cost function of producing x hammers is $C(x) = 4x + 170$. If each hammer is sold for $10, the revenue function for selling x hammers is $R(x) = 10x$.
How many hammers must be sold to make a profit?

How many hammers must be sold to make a profit of $100?

Basic Circle Formulas

A quick review of circle formulas and basic circle properties:

Area of a circle: $$\pi r^2$$

Circumference: $$\pi d \text{ or } 2\pi r$$

We will also need to know that the degree measure of a central angle is equal to the degree measure of the arc it intercepts.

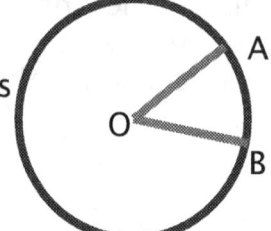

For example, if the measure of angle AOC is 35° then the measure of minor arc (the smaller arc connecting points A and B) is also 35°.

What is the measure of major arc AB? (The measure of the larger arc connecting points A and B.) _____

A new formula that is very handy with relationships between circle parts is

$$\theta = \frac{s}{r}$$

where "θ" (theta) is the measure of a central angle in radians (we'll discuss what a radian is later), "s" is the measure of the intercepted arc in the same units as the radius is measured in, and "r" is the measure of the radius.

Sample #32: If an arc of 60° on circle A has the same length as an arc of 45° on circle B, what is the ratio of the area of circle B to the area of circle A?

June '01, #6: The circumference of a circular plot of land is increased by 10%. What is the best estimate of the total percentage that the area of the plot is increased?

(1) 10%   (2) 21%   (3) 25%   (4) 31%

Jan '02, #15: Every time the pedals go through a 360° rotation on a certain bicycle, the tires rotate three times. If the tires are 24 inches in diameter, what is the minimum number of complete rotations of the pedals needed for the bicycle to travel at least 1 mile?

(1) 12   (2) 281   (3) 561   (4) 5280

Jan '02, #23: A ball is rolling in a circular path that has a radius of 10 inches, as shown in the accompanying diagram. What distance has the ball rolled when the subtended arc is 54°? Express your answer to the nearest hundredth of an inch.

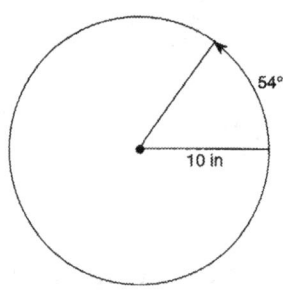

Jan '03, #7: Ileana buys a large circular pizza that is divided into eight equal slices. She measures along the outer edge of the crust from one piece and finds it to be $5\frac{1}{2}$ inches. What is the diameter of the pizza to the nearest inch?

(1) 14   (2) 8   (3) 7   (4) 4

Jan '04, #3: An overhead view of a revolving door is shown in the accompanying diagram. Each panel is 1.5 meters wide.

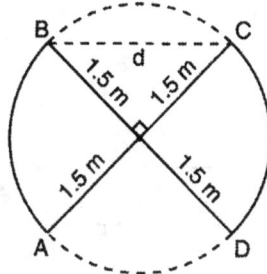

What is the approximate width of d, the opening from B to C?
(1) 1.50 m   (2) 1.73 m   (3) 3.00 m   (4) 2.12 m

Parametric Equations

Some equations that we cannot graph in function mode can be graphed in **_parametric mode_**.

Give 3 examples of equations that cannot be graphed in function mode:

1. _____

2. _____

3. _____

Review: What is a function?

_____

_____

How can we identify a function from an equation?

_____

_____

How can we identify a function from a graph?

_____

_____

Some equations can be redefined as two functions where each function represents the x or y of the original equation and are functions of a third variable usually called T.

Let's begin with a basic example:

The unit circle

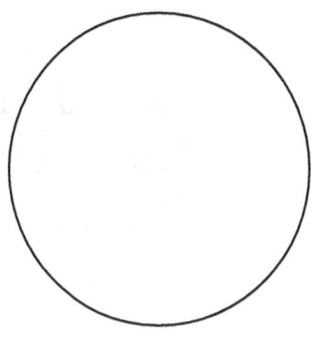

The unit circle is:

It is basic to the study of trigonometry. Therefore, it should make sense that we can define this circle in terms of two trigonometric functions.

X will be represented by *cos(T)*
Y will be represented by *sin(T)*

Then our ordered pairs will be _____

with _____ ≤ T ≤ _____

Verify that *cos(T)* and *sin(T)* are functions:

1. Begin by checking to see that your calculator is in **radian mode**.

2. Graph *cos(x)* using ZoomTrig and sketch below.

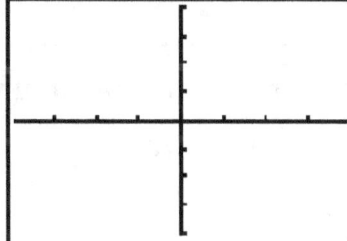

3. Graph *sin(x)* and sketch below.

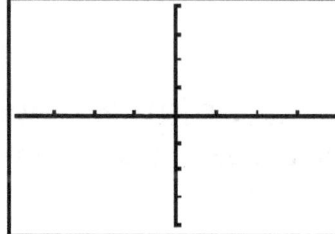

4. What justification can be given to say that *sin(x)* and *cos(x)* are functions?

The window will need to be adjusted.

Radians ◄ · – ► Degrees

First we will explore the connection between degrees and radians in the unit circle.

1. How many degrees are in a full rotation of the unit circle?

----

2. What is the circumference of the unit circle?

----

3. In a circle, the measure of a central angle is equal to

----

4. Describe the relationship between the circumference of the unit circle and a full rotation of the circle.

Bringing all this together, a **radian** is how large a central angle must be to intercept an arc that is exactly the same length as the radius of the circle.

There are formulas to convert radians to degrees and visa versa but we will look at a graphing calculator method instead.

Recall that until now we have always changed the MODE to DEGREES before working with trigonometric functions. We will need to pay closer attention now, as we will sometimes want DEGREE mode and sometimes want RADIAN mode.

To convert an angle measure from degrees to radians:
1. Be sure that the calculator's mode is that of the answer, in this case radians.
2. Enter the degrees followed by the degree symbol (°) from the angle menu.
3. Press ENTER.

If a decimal answer is sufficient you are finished. Often radians are written as a fraction in terms of π. If this type of answer is required continue with the steps below.
4. Divide by π. (Be sure to use the π key, not an approximation for π.)
5. Convert the remaining decimal to a fraction.
6. Write the fraction with "π" included in the numerator.

To convert an angle measure from radians to degrees:
1. Be sure that the calculator's mode is in degrees.
2. Enter the radians followed by "r". This lower case r is choice #3 in the angle menu.
3. Press ENTER.

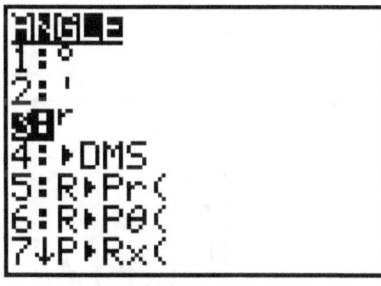

Usually the decimal form is appropriate, but if degrees, minutes, seconds form is desired, return to the angle menu and choose #4, then ENTER.

***If the radian measure is in fraction form, be sure to use parentheses around the angle measure with the "r" outside the parentheses.

Practice:

Convert the following degree measures to radians. Give both the decimal and fraction forms.

1. 30°   _____   _____

2. 45°   _____   _____

3. 60°   _____   _____

4. 90°   _____   _____

5. 125°   _____   _____

6. 150°   _____   _____

7. 175°   _____   _____

8. 190°   _____   _____

9. 220°   _____   _____

10. 310°   _____   _____

Convert the following radian measures to degrees. If the answer is a decimal give the decimal to the nearest hundredth and convert to degrees, minutes, and seconds and give the answer to the nearest second.

11. $\dfrac{\pi}{5}$ _____

12. $\dfrac{2\pi}{9}$ _____

13. $\dfrac{3\pi}{4}$ _____

14. $\dfrac{3\pi}{11}$ _____   _____

15. $\dfrac{5\pi}{8}$ _____   _____

16. $\dfrac{7\pi}{16}$ _____   _____

17. $\dfrac{4\pi}{13}$ _____   _____

Back to parametric graphing:

List the WINDOW dimensions given by ZoomTrig:

Xmin= _____

Xmax= _____

Xscl= _____

Ymin= _____

Ymax= _____

Yscl= _____

Change your mode from function (Func) to parametric (Par).

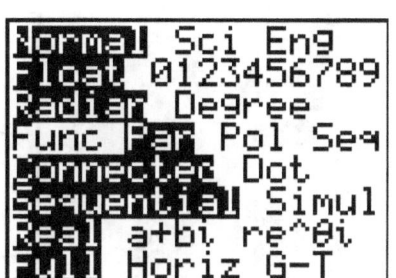

Go to the WINDOW. You should notice some changes.
Adjust the settings to the following:

Tmin=0
Tmax=2π
Tstep=.1
Xmin=-1.5
Xmax=1.5
Xscl=π/2
Ymin=-1.25
Ymax=1.25
Yscl=π/2

Press Y=.

The screen below should appear.

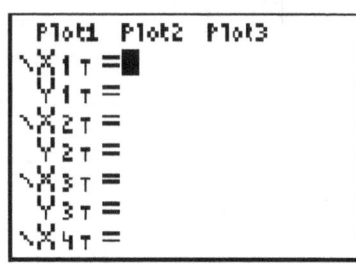

For each graph you will need to identify 2 functions, the X function and the Y function.

Enter *cos(T)* in X_1.
Enter *sin(T)* in Y_1.

****Notice that the key we have always used for "x" now enters "T" on our screen.****

Graph and sketch.

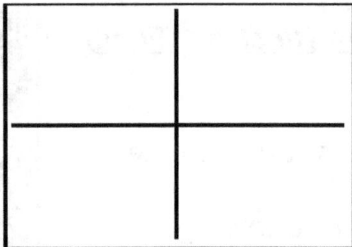

Press TRACE.

Enter the following values for *T* and record the resulting x and y values. (Change Tmax to 6.3)

T	X	Y
0		
$\frac{\pi}{2}$		
π		
$\frac{3\pi}{2}$		
2π		

Unit Circle and Radian Measure

<u>Aug '02, #23:</u> An art student wants to make a string collage by connecting six equally spaced points on the circumference of a circle to its center with string. What would be the radian measure of the angle between two adjacent pieces of string, in simplest form?

<u>Aug '01, #9:</u> A regular hexagon is inscribed in a circle. What is the ratio of the length of a side of the hexagon to the minor arc that it intercepts?

(1) $\dfrac{\pi}{6}$ (2) $\dfrac{3}{6}$ (3) $\dfrac{3}{\pi}$ (4) $\dfrac{6}{\pi}$

<u>Aug '01, #16:</u> A wedge-shaped piece is cut from a circular pizza. The radius of the pizza is 6 inches. The rounded edge of the crust of the piece measures 4.2 inches. To the nearest tenth, the angle of the pointed end of the piece of pizza, in radians, is

(1) 0.7   (2) 1.4   (3) 7.0   (4) 25.2

<u>Jan '02, #5:</u> If θ is an angle in standard position and its terminal side passes through the point $\left(\dfrac{1}{2}, \dfrac{\sqrt{3}}{2}\right)$ on a unit circle, a possible value of θ is

(1) 30°   (2) 60°   (3) 120°   (4) 150°

June '02, #3: In the accompanying diagram, the length of arc ABC is $\frac{3\pi}{2}$ radians.

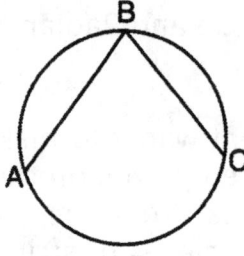

(Not drawn to scale)

What is m∠ABC?

(1) 36   (2) 45   (3) 53   (4) 72

June '01, #20: Through how many radians does the minute hand of a clock turn in 24 minutes?

(1) 0.2π   (2) 0.4π   (3) 0.6π   (4) 0.8π

Sample #20: The origin of a coordinate grid is labeled A. Line segment AB forms an angle of 30° with the x-axis. If AB=8, the coordinates of B are:

(1) (6,4)   (3) *(8sin30°,8cos30°)*
(2) *(8cos30°,8sin30°)*   (4) *(4,4√3)*

June '03, #2: If *sin θ >0* and *sec θ <0*, in which quadrant does the terminal side of angle θ lie?

(1) I   (2) II   (3) III   (4) IV

<u>Aug '03, #9:</u> A dog has a 20-foot leash attached to the corner where a garage and a fence meet, as shown in the accompanying diagram. When the dog pulls the leash tight and walks from the fence to the garage, the arc the leash makes is 55.8 feet.

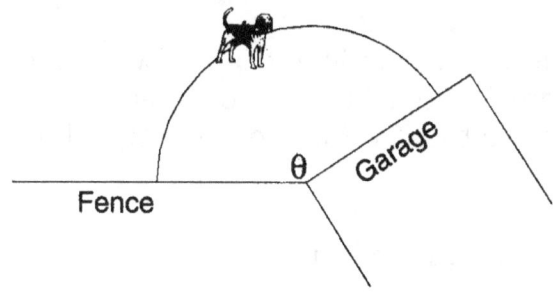

(Not drawn to scale)

What is the measure of angle θ between the garage and the fence, in radians?

(1) 0.36   (2) 2.79   (3) 3.14   (4) 160

<u>Jan '04, #21:</u> Kristine is riding in car 4 of the Ferris wheel represented in the accompanying diagram. The Ferris wheel is rotating in the direction indicated by the arrows. The eight cars are equally spaced around the circular wheel. Express, in radians, the measure of the smallest angle through which she will travel to reach the bottom of the Ferris wheel.

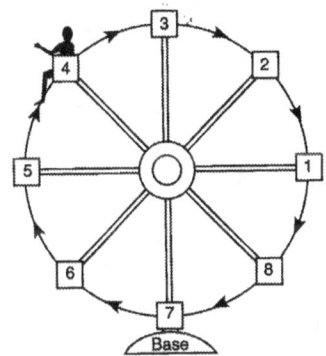

<u>Jan '04 #22:</u> In the accompanying diagram, point P(0.6, -0.8) is on unit circle O. What is the value of θ, to the nearest degree?

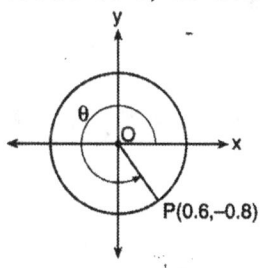

More Parametric Graphing

Begin with calculator in radian mode and window settings at:
Zoom Standard then Zoom Decimal
Change the T values to the interval given.

Circles:

A circle with center (c,d) and radius r is entered:

----

with $0 \leq T \leq 2\pi$

Parabolas:

With horizontal axis is entered:

----

with $-2\pi \leq T \leq 2\pi$

With vertical axis is entered:

----

with $-2\pi \leq T \leq 2\pi$

***Always set ellipses and non-rectangular hyperbolas equal to 1 first!!!**

Ellipses:
  Entered as:

  Where a= ½ the length of the horizontal axis or the square root
  of the "*x*" denominator when the equation is set equal to 1.

  And b= ½ the length of the vertical axis of the square root
  of the "*y*" denominator when the equation is set equal to 1.
  With $0 \leq T \leq 2\pi$

Hyperbolas: (non-rectangular)
  When *x* is part of the "positive" fraction:

  When *x* is part of the "negative" fraction:

  ----

  where a=the square root of the positive denominator
  b=the square root of the negative denominator
  with $0 \leq T \leq 2\pi$

Try this one:

  X=4cos(T)+cos(3T)
  Y=sin(3T)+T

  with $0 \leq T \leq 4\pi$
  change Ymax to 10
  (change to ZoomTrig first)

Practice:
Graph each equation in parametric mode. Write the name of the shape on the space provided and sketch. Be sure the x and y intercepts are accurate!

1. $y = x^2 - 3x + 2$ _____

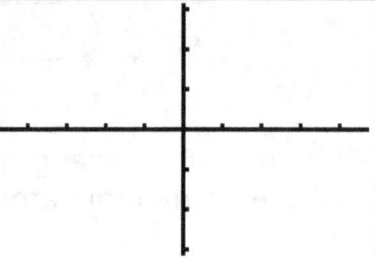

        Window: Zoom Decimal
$$-2\pi \le T \le 2\pi$$

2. $y = -2x^2 + 5x + 7$ _____

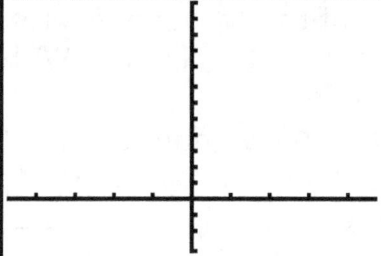

        Window: Change Ymax to 12

3. $x = 2y^2 + 3$

      _____

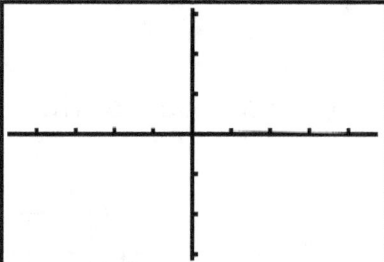

        Window: Return Ymax to 3.1

4. $x = -3y^2 + y + 2$ _____

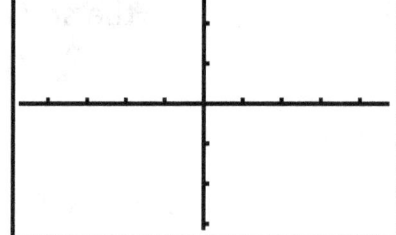

5. $x^2 + y^2 = 16$ _____
        Window: Zoom Standard
           then Zoom Square

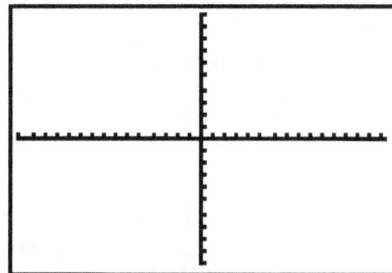

6. $(x-3)^2 + (y+7)^2 = 9$ _____

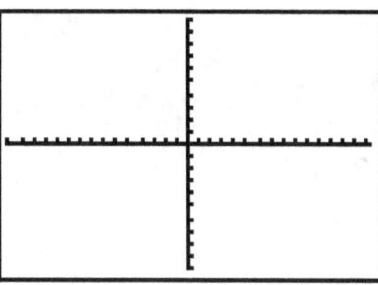

7. $(x+2)^2 + (y-3)^2 = 4$ _____

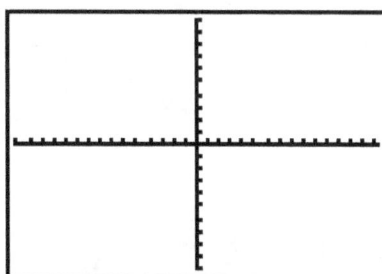

8. $\dfrac{x^2}{4} + \dfrac{y^2}{16} = 1$ _____

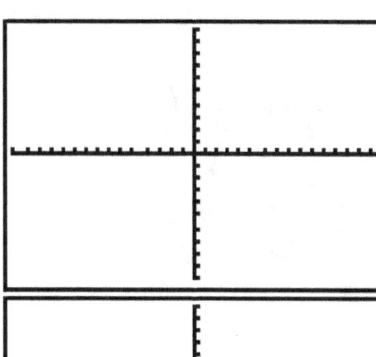

9. $4x^2 + 25y^2 = 100$ _____

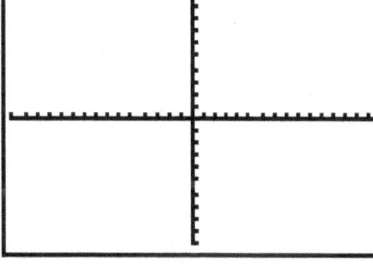

10. $\dfrac{x^2}{9} - \dfrac{y^2}{16} = 1$ _____

    **Window: Change to Degree Mode
        then Zoom Standard
        then Zoom Decimal**

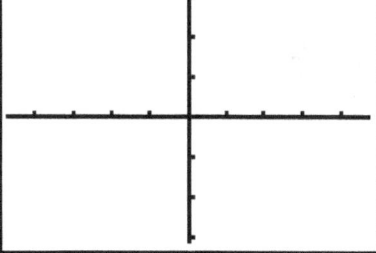

11. $\dfrac{y^2}{25} - \dfrac{x^2}{9} = 1$ _____

    **Window: Zoom Standard
        then Zoom Square**

12. $4x^2 + 9y^2 = 36$ _____

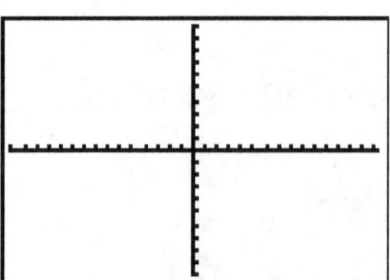

13. $9x^2 - 4y^2 = 36$ _____

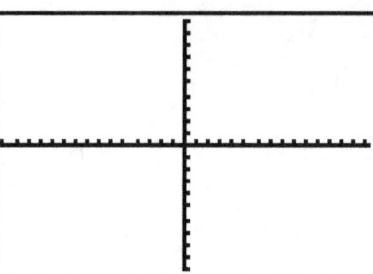

14. $\dfrac{x^2}{49} + \dfrac{y^2}{100} = 1$ _____

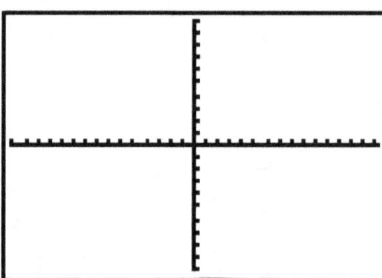

15. $16y^2 - 4x^2 = 64$ _____

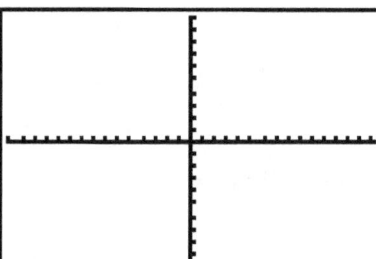

16. $\dfrac{y^2}{25} - x^2 = 1$ _____

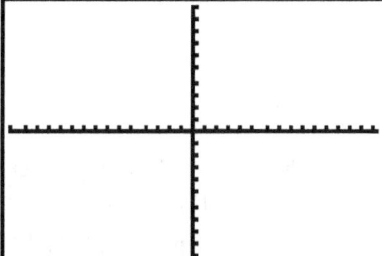

17. $9x^2 - 16y^2 = 144$ _____

Conic Sections

<u>Sample #17:</u> The center and radius of the given circle
$(x-3)^2 + (y+8)^2 = 39$ are:

    (1) (3,-8), $r=39$ (3) (-3,8), $r=\sqrt{39}$

    (2) (-3,-8), $r=\sqrt{39}$ (4) (3,-8), $r=\sqrt{39}$

<u>June '01, #10:</u> The center of a circular sunflower with a diameter of 4 centimeters is (-2,1). Which equation represents the sunflower?

    (1) $(x-2)^2 + (y+1)^2 = 2$

    (2) $(x+2)^2 + (y-1)^2 = 4$

    (3) $(x-2)^2 + (y-1)^2 = 4$

    (4) $(x+2)^2 + (y-1)^2 = 2$

<u>June '03, #11:</u> The accompanying diagram represents the elliptical path of a ride at an amusement park.

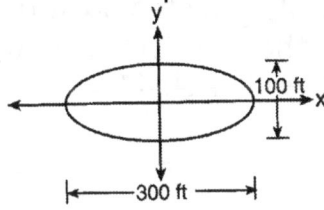

Which equation represents this path?

    (1) $x^2 + y^2 = 300$ (3) $\dfrac{x^2}{150^2} + \dfrac{y^2}{50^2} = 1$

    (2) $y = x^2 + 100x + 300$ (4) $\dfrac{x^2}{150^2} - \dfrac{y^2}{50^2} = 1$

June '03, #34: For a carnival game, John is painting two circles, V and M, on a square dartboard.

a. On the accompanying grid, draw and label circle V, represented by the equation $x^2 + y^2 = 25$, and circle M, represented by the equation $(x-8)^2 + (y+6)^2 = 4$.

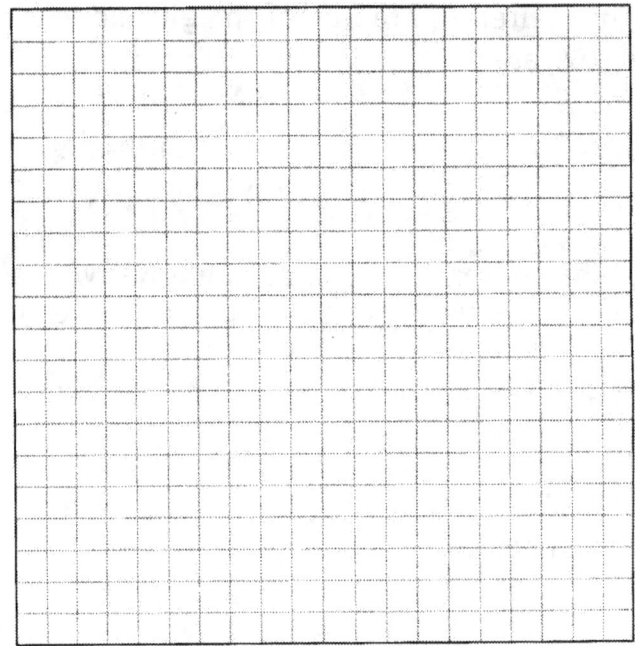

b. A point, (x,y), is randomly selected such that $-10 \le x \le 10$ and $-10 \le y \le 10$. What is the probability that point (x,y) lies outside both circle V and circle M?

<u>Aug '02, #6:</u> An architect is designing a building to include an arch in the shape of a semi-ellipse (half an ellipse), such that the width of the arch is 20 feet and the height of the arch is 8 feet, as shown in the accompanying diagram.

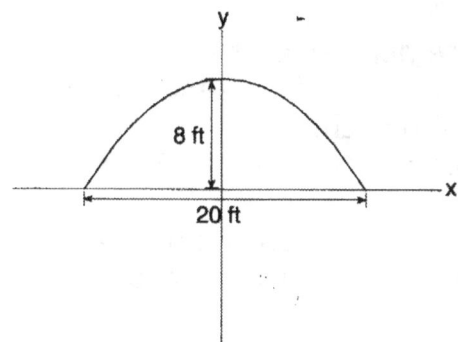

Which equation models this arch?

(1) $\dfrac{x^2}{100}+\dfrac{y^2}{64}=1$ (3) $\dfrac{x^2}{64}+\dfrac{y^2}{100}=1$

(2) $\dfrac{x^2}{400}+\dfrac{y^2}{64}=1$ (4) $\dfrac{x^2}{64}+\dfrac{y^2}{400}=1$

<u>Aug '03, #18:</u> A commercial artist plans to include an ellipse in a design and wants the length of the horizontal axis to equal 10 and the length of the vertical axis to equal 6. Which equation could represent this ellipse?

(1) $9x^2 + 25y^2 = 225$ (3) $x^2 + y^2 = 100$

(2) $9x^2 - 25y^2 = 225$ (4) $3y = 20x^2$

<u>Jan '04, #10:</u> The accompanying diagram shows the elliptical orbit of a planet. The foci of the elliptical orbit are F_1 and F_2

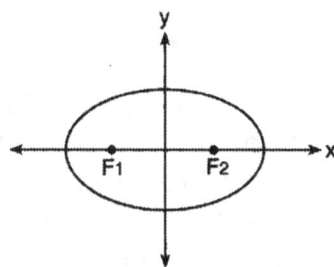

If a, b, and c are all positive and $a \neq b \neq c$, which equation could represent the path of the planet?

     (1) $ax^2 - by^2 = c^2$ (3) $y = ax^2 + c^2$

     (2) $ax^2 + by^2 = c^2$ (4) $x^2 + y^2 = c^2$

Inverse Variation

<u>Aug '01, #23:</u> The price per person to rent a limousine for a prom varies inversely as the number of passengers. If five people rent the limousine, the cost is $70 each. How many people are renting the limousine when the cost per couple is $87.50?

<u>Jan '02, #21:</u> Explain how a person can determine if a set of data represents inverse variation and give an example using a table of values.

<u>June '01, #4:</u> Camisha is paying a band $330 to play at her graduation party. The amount each member earns, d, varies inversely as the number of members who play, n. The graph of the equation that represents the relationship between d and n is an example of

(1) a hyperbola   (2) a line   (3) a parabola   (4) an ellipse

<u>June '03, #23:</u> When air is pumped into an automobile tire, the pressure is inversely proportional to the volume. If the pressure is 35 pounds when the volume is 120 cubic inches, what is the pressure, in pounds, when the volume is 140 cubic inches?

<u>Jan '03, #10:</u> For a rectangular garden with a fixed area, the length of the garden varies inversely with the width. Which equation represents this situation for an area of 36 square units?

(1) $x + y = 36$ (2) $y = \dfrac{36}{x}$ (3) $x - y = 36$ (4) $y = 36x$

<u>Jan '04, #23:</u> A pulley that has a diameter of 8 inches is belted to a pulley that has a diameter of 12 inches. The 8-inch-diameter pulley is running at 1,548 revolutions per minute. If the speeds of the pulleys vary inversely to their diameters, how many revolutions per minute does the larger pulley make?

Exploring Trigonometric Curves

Let's start by graphing the basic sine, cosine, and tangent curves. Enter the given functions in Y$_1$ one at a time and sketch in the space provided. For consistency, be sure you are in degree mode and choose Zoom Standard, then Zoom Trig before beginning the trigonometric graphs.

$y = \sin x$

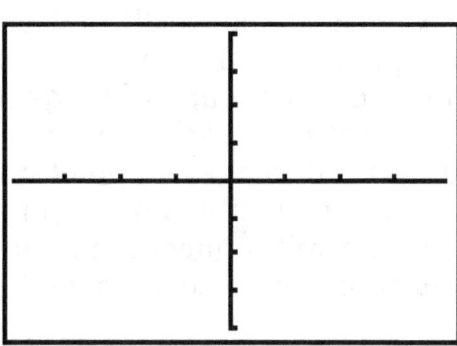

$y = \cos x$

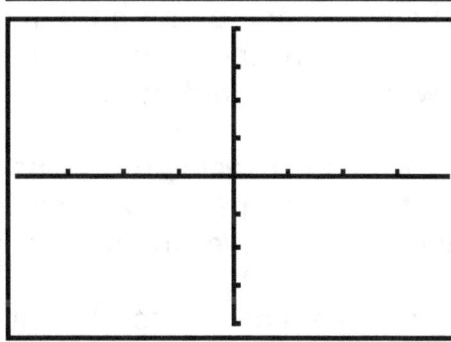

$y = \tan x$

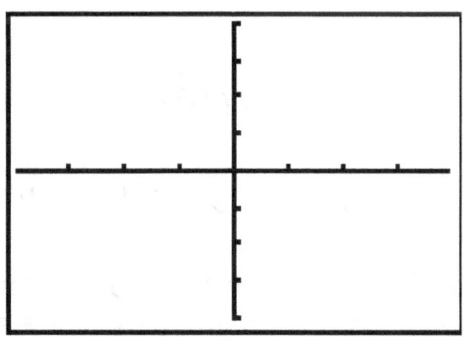

List the current window dimensions below:

Xmin: _____   Ymin: _____

Xmax: _____   Ymax: _____

Xscl: _____   Yscl: _____

Change the minimum x to –360 and the maximum x to 360.
Graph $y = \tan x$ again and
sketch.

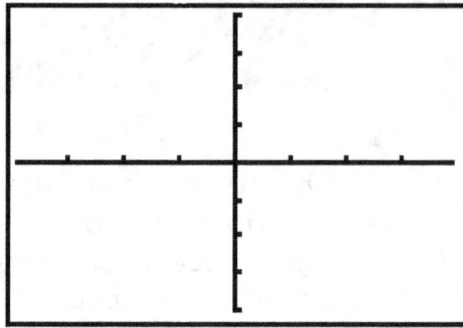

These new lines are not really part of the graph of tangent x. These actually approximate the asymptotes. (The values of x where tangent is undefined but the graph of tangent will come infinitely close to these values without reaching them.) The lines appear because the calculator doesn't plot every point on a graph. It plots many points and connects them. If it is not trying to plot the points where tangent is undefined, it will connect the points on either side creating the graph you should have seen on your screen.

It won't be necessary to graph tangent very often so we won't explore this function any further. Just make note that it will be best to graph tangent with Zoom Trig if possible.

Just like the absolute value functions we graphed previously, changes in different parts of the sine and cosine functions will change the appearance of the curves in different ways.

Let's begin by naming the coefficients and constants.

$$y = a \sin bx + c \quad \text{and}$$
$$y = a \cos bx + c$$

Graph and sketch each function on the following pages. Compare the graphs with the basic graphs of sine and cosine and write your conclusions about the effects of "a", "b", and "c" in the space provided.

1. $y = 2\sin x$

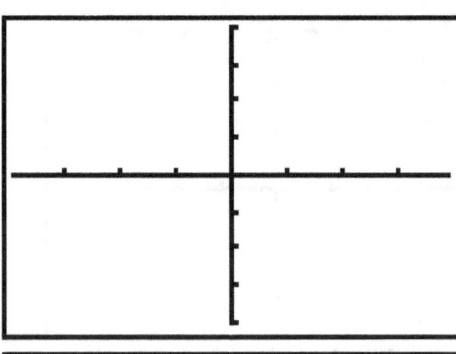

2. $y = \dfrac{1}{2}\sin x$

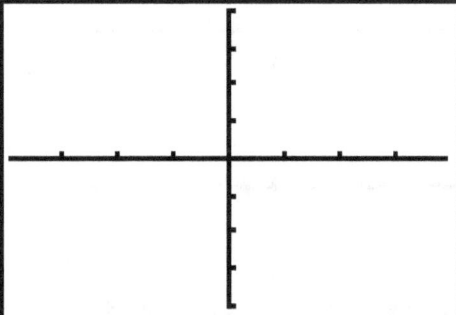

3. $y = 4\sin x$

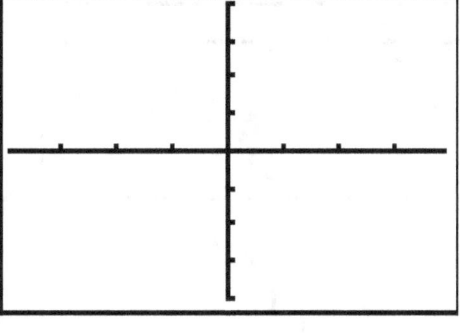

What conclusion can you make about the effect that "a" has on the appearance of the graph of the sine or cosine equation?

_____

_____

_____

_____

4. $y = 3\cos x$

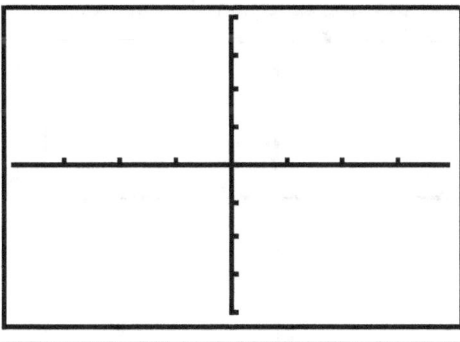
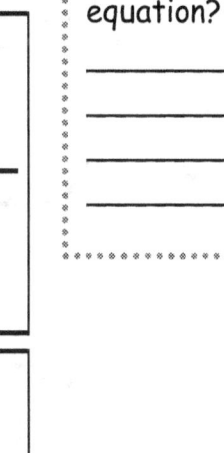

5. $y = \dfrac{2}{5}\cos x$

6. $y = \sin 3x$

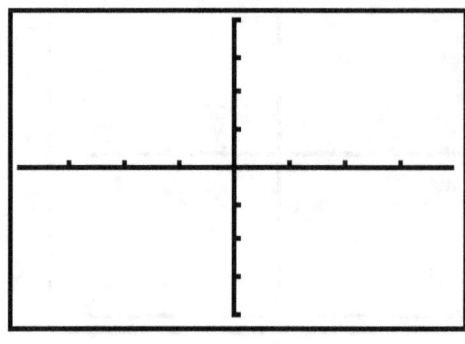

7. $y = \sin \dfrac{1}{2} x$

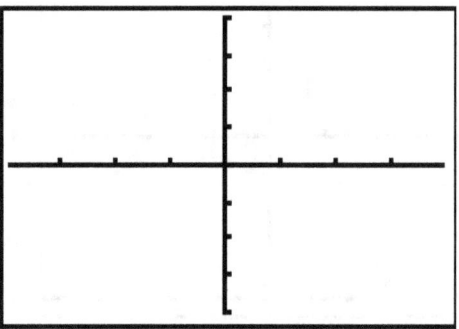

8. $y = \cos 5x$

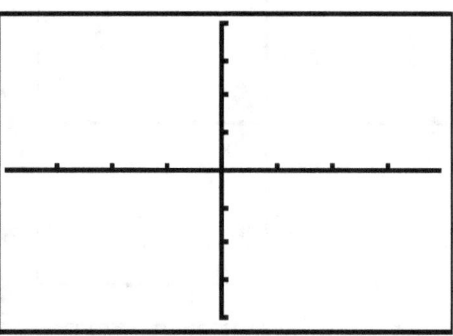

What conclusion can you make about the effect that "b" has on the appearance of the graph of the sine or cosine equation?

_____
_____
_____
_____

9. $y = \cos \dfrac{2}{3} x$

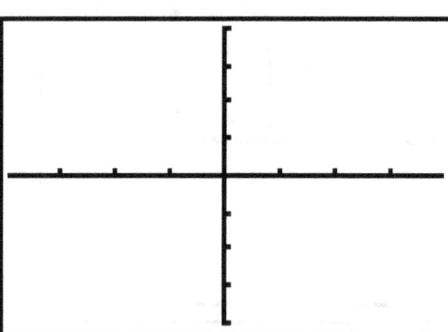

10. $y = \cos 10x$

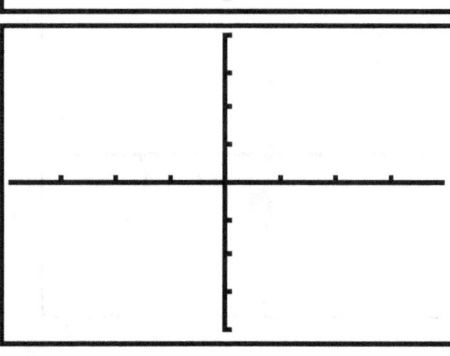

11. $y = \sin x + 2$

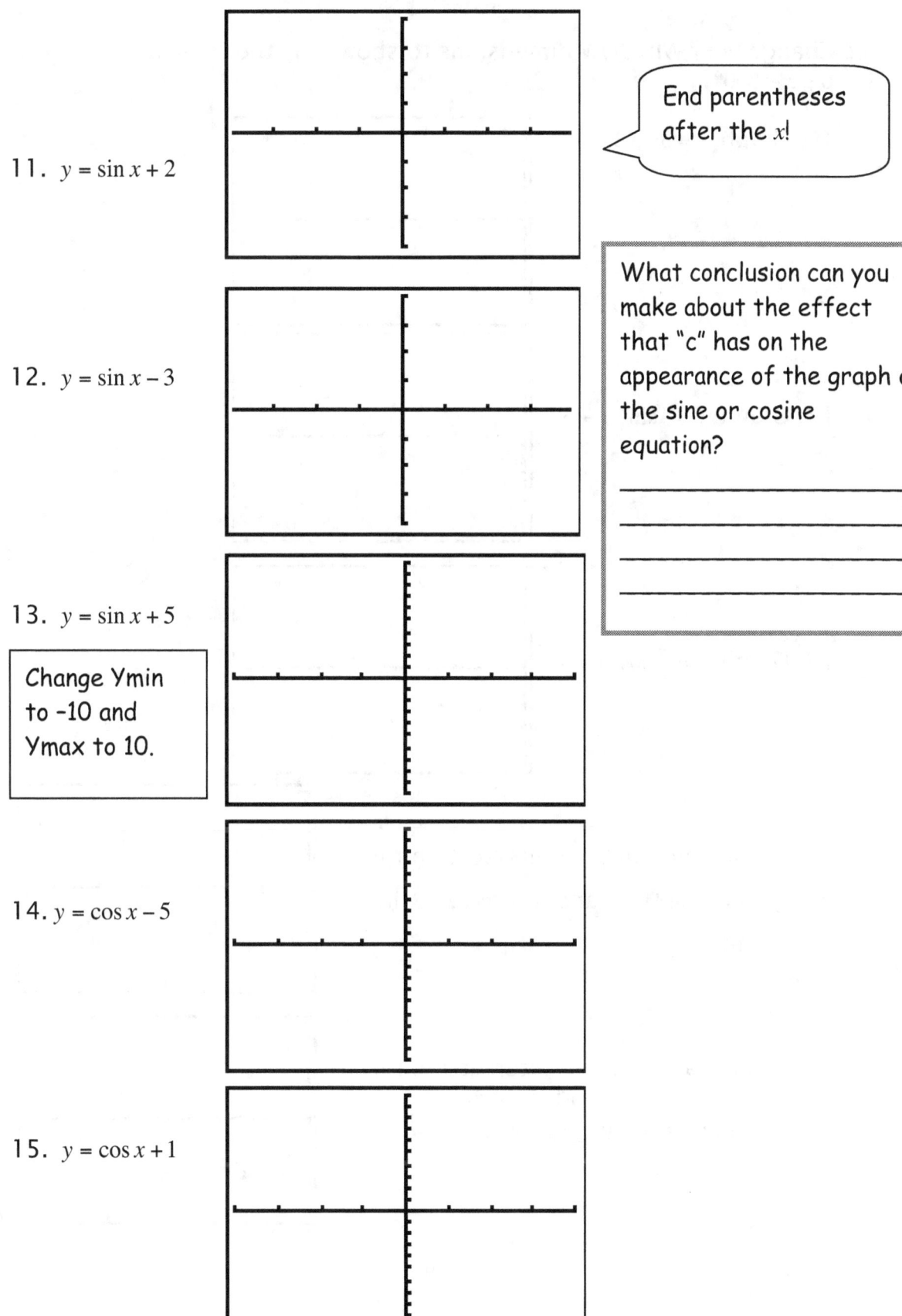

End parentheses after the x!

12. $y = \sin x - 3$

What conclusion can you make about the effect that "c" has on the appearance of the graph of the sine or cosine equation?

_____
_____
_____
_____

13. $y = \sin x + 5$

Change Ymin to –10 and Ymax to 10.

14. $y = \cos x - 5$

15. $y = \cos x + 1$

Change your WINDOW dimensions to show only the interval $0 \leq x \leq 360°$.

16. Graph $y = 3\sin 2x - 4$

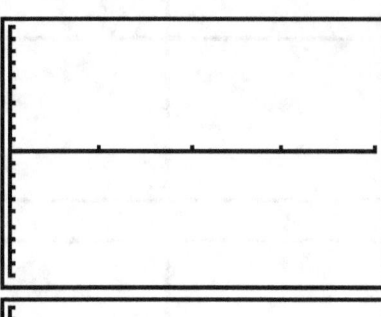

17. Graph $y = 4\sin\dfrac{1}{3}x - 3$

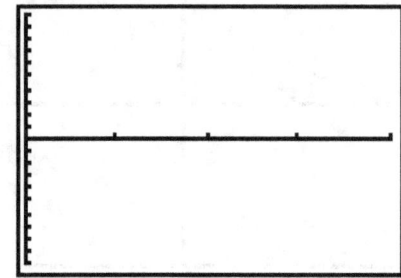

18. Graph $y = \dfrac{4}{5}\cos 3x + 5$

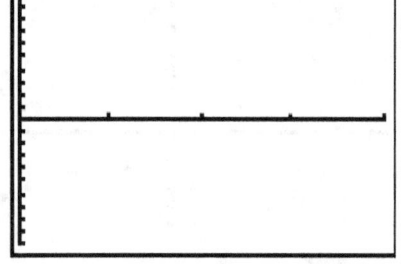

19. Find all points of intersection for $y = \dfrac{1}{2}\sin 3x + 4$ and $y = 2\cos x + 3$ in the interval $0 \leq x \leq 360°$.

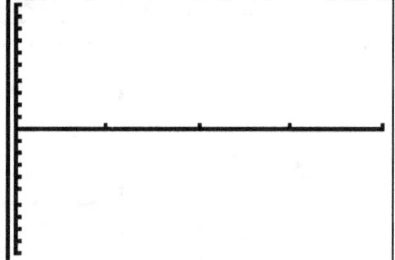

20. Find all points of intersection for $y = 3\sin 2x - 5$ and $y = 4\cos 3x + 1$ in the interval $0 \leq x \leq 360°$.

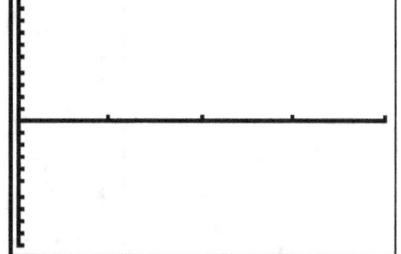

a ~ b ~ c

"a" is called the _____

"b" is called the _____

"c" is called the _____

From the frequency we can also find the "period" of the curve. The period is

_____

_____

Graph $y = \sin 3x$ with the window showing the interval $0 \leq x \leq 360°$, $-4 \leq y \leq 4$.

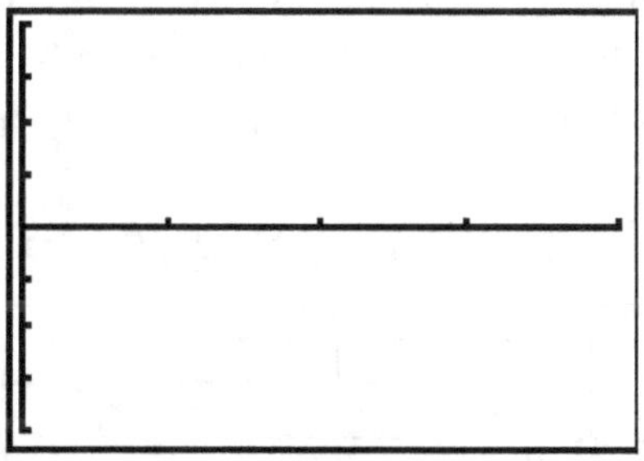

How many times does the curve "repeat" in this interval? _____
How many degrees are in each copy of the curve? _____

To find the period of any trigonometric function divide the number of degrees or the number of radians in a full circle by the frequency.

Amplitude, Period, Trigonometric Graphs

<u>Aug '01, #13:</u> What is the period of the function $y = 5\sin 3x$?

   (1) 5   (2) $\dfrac{2\pi}{5}$ (3) 3   (4) $\dfrac{2\pi}{3}$

<u>Aug '01, #21:</u> If the sine of an angle is $\dfrac{3}{5}$ and the angle is not in Quadrant I, what is the value of the cosine of the angle?

<u>Aug '01, #26:</u> If $\sin x = \dfrac{4}{5}$, where $0° < x < 90°$, find the value of $\cos(x + 180°)$.

<u>Aug '01, #27:</u> The times of average monthly sunrise, as shown in the accompanying diagram, over the course of a 12-month interval can be modeled by the equation $y = A\cos(Bx) + D$. Determine the values of *A*, *B*, and *D*, and explain how you arrived at your answer.

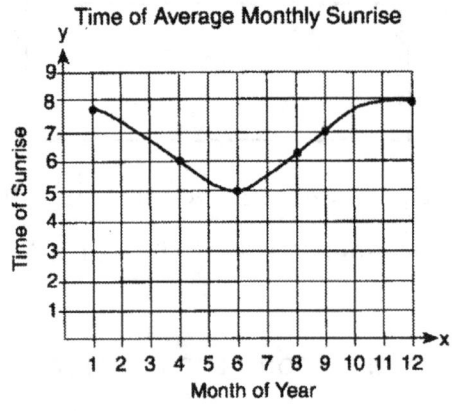

<u>June '01, #5:</u> A modulated laser heats a diamond. Its variable temperature, in degrees Celsius, is given by $f(t) = T\sin a$. What is the period of the curve?

   (1) $|T|$ (2) $\dfrac{2\pi}{a}$ (3) $\dfrac{1}{a}$ (4) $\dfrac{2a\pi}{a}$

Jan '02, #14: The accompanying diagram shows a section of a sound wave as displayed on an oscilloscope.

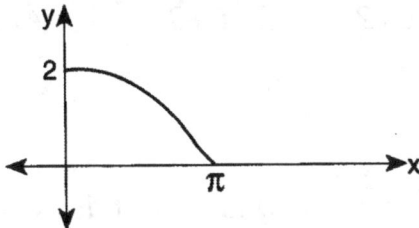

Which equation could represent this graph?

(1) $y = 2\cos\dfrac{x}{2}$ (2) $y = 2\sin\dfrac{x}{2}$ (3) $y = \dfrac{1}{2}\cos 2x$ (4) $y = \dfrac{1}{2}\sin\dfrac{\pi}{2}x$

Jan '02, #16: Which type of symmetry does the equation $y = \cos x$ have?

(1) line symmetry with respect to the x-axis
(2) line symmetry with respect to the line y=x
(3) point symmetry with respect to the origin
(4) point symmetry with respect to $\left(\dfrac{\pi}{2}, 0\right)$

Sample #31: A helicopter, starting at point A on Sunrise Highway, circles a 2-mile section of the highway in a counterclockwise direction. If the helicopter is traveling at a constant speed and it takes approximately 6.28 minutes to make one complete revolution to return to point A, sketch a possible graph of distance (dependent variable) from the helicopter to the highway, versus time (independent variable). If the helicopter is north of the highway, distance (d) is positive; if the helicopter is south of the highway, distance (d) is negative. (Disregard the height of the helicopter.) State the equation of this graph.

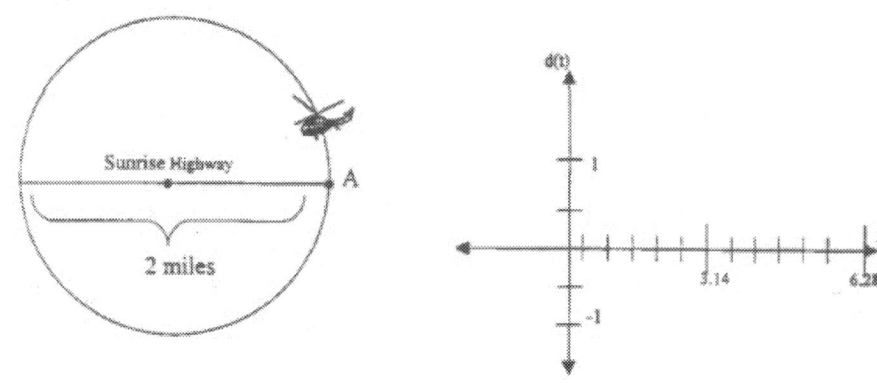

<u>Sample #19</u>: If $f(x) = 2\sin x + C$, then the maximum value of $f(x)$ is:

(1) C (2) $C+2$ (3) $C+3$ (4) $C+6$

<u>Jan '02, #4</u>: An object that weighs 2 pounds is suspended in a liquid. When the object is depressed 3 feet from its equilibrium point, it will oscillate according to the formula $x = 3\cos(8t)$, where t is the number of seconds after the object is released. How many seconds are in the period of oscillation?

(1) $\dfrac{\pi}{4}$ (2) π (3) 3   (4) 2π

<u>June '03, #29</u>: A pair of figure skaters graphed part of their routine on a grid. The male skater's path is represented by the equation $m(x) = 3\sin\dfrac{1}{2}x$, and the female skater's path is represented by the equation $f(x) = -2\cos x$. On the accompanying grid, sketch both paths and state how many times the paths of the skaters intersect between $x=0$ and $x=4\pi$.

<u>Jan '03, #1:</u> A monitor displays the graph $y = 3\sin 5x$. What will be the amplitude after a dilation of 2?

    (1) 5   (2) 6   (3) 7   (4) 10

<u>Jan '03, #29:</u> A building's temperature, T, varies with time of day, t, during the course of 1 day, as follows:

$$T = 8\cos t + 78$$

The air-conditioning operates when $T \geq 80°F$. Graph this function for $6 \leq t \leq 17$ and determine, to the nearest tenth of an hour, the amount of time in 1 day that the air-conditioning is on in the building.

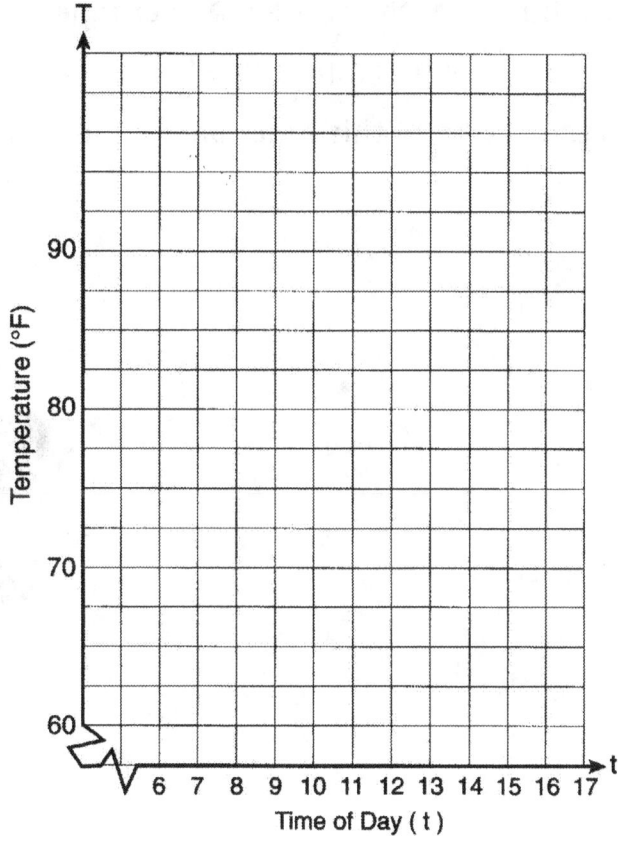

<u>Aug '03, #30:</u> A student attaches one end of a rope to a wall at a fixed point 3 feet above the ground, as shown in the accompanying diagram, and moves the other end of the rope up and down, producing a wave described by the equation $y = a \sin bx + c$. The range of the rope's height above the ground is between 1 and 5 feet. The period of the wave is 4π. Write the equation that represents this wave.

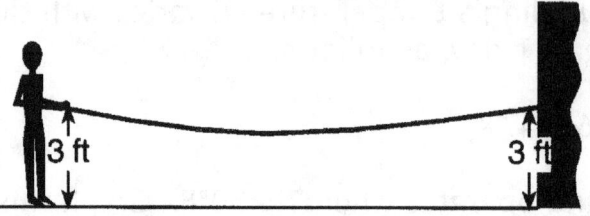

<u>Jan '04, #25:</u> The brightness of the star MIRA over time is given by the equation $y = 2\sin\dfrac{\pi}{4} + 6$, where x represents time and y represents brightness. What is the period of this function in radian measure?

Other Regressions:
Exponential, Logarithmic, Power

The four types of regression we are expected to study for Math B are linear, exponential, logarithmic, and power.

We have already looked at linear and power regressions. Linear are the most common so they were covered in depth earlier and won't be repeated here.

The other types are similar and are started the same way:
1. Go to the CATALOG (2nd ,0).
2. Find Diagnostics On. Press ENTER twice. The home screen should say Done.
3. Enter the data for the *x*-axis (the independent variable) in L_1.
4. Enter the data for the *y*-axis (the dependent variable) in L_2.
5. Press STAT.
6. Go over to CALC.
7. Choose the type of regression.
    a. For linear choose
        4:LinReg(ax+b)
    b. For Exponential choose
        0:ExpReg
    c. For Power choose A:PwrReg
    d. For Logarithmic choose
        9:LnReg
8. On the home screen press ENTER.

    (Unless the data is in lists other than

    L_1 and L_2, then the lists need to be

    named.)

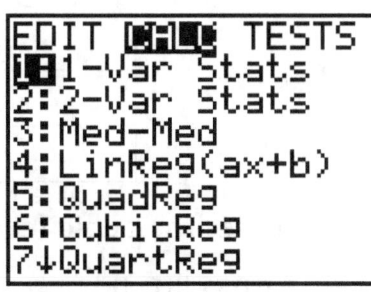

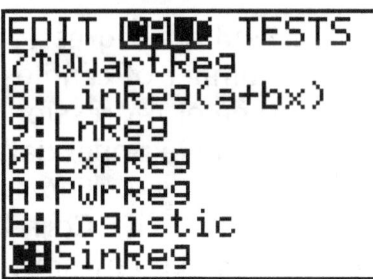

Because the three types of equations can behave similarly in the first quadrant, where regression data usually lies, we will use the same data for each type and see which is a "better fit".

The data below describes the number of candles needed to power different sized candle-powered hot-air balloons at 69°F.

Cubic Feet (Size of Balloon)	Number of Candles	Gross Lift (in ounces)
1	7	0.25
5	20	1.25
10	32	2.50
15	42	3.75
20	50	5.00
25	58	6.25
30	66	7.50
35	73	8.75
40	80	10.00
45	86	11.25
50	93	12.50

Source: http://www.overflight.com/power.html

1. Enter the cubic feet in L$_1$, the number of candles in L$_2$, and the gross lift in L$_3$.

2. Find the regression equation comparing cubic feet (x-list) with number of candles (y-list) and correlation coefficient for each type of regression below:
   a. Exponential
      i. Equation: _____
      ii. Correlation Coefficient: _____
   b. Logarithmic
      i. Equation: _____
      ii. Correlation Coefficient: _____
   c. Power
      i. Equation: _____
      ii. Correlation Coefficient: _____
   d. Which equation has a better correlation? _____
      Explain. _____

3. Find the regression equation comparing cubic feet (*x*-list) with gross lift (*y*-list) and correlation coefficient for each type of regression below: (Use L_1, L_3 after the regression type.)

    a. Exponential
        i. Equation: _____
        ii. Correlation Coefficient: _____
    b. Logarithmic
        i. Equation: _____
        ii. Correlation Coefficient: _____
    c. Power
        i. Equation: _____
        ii. Correlation Coefficient: _____
    d. Which equation has a better correlation? _____

    Explain. _____

4. Find the regression equation comparing number of candles (*x*-list) with gross lift (*y*-list) and correlation coefficient for each type of regression below: (Use L_2, L_3 after the regression type.)

    a. Exponential
        i. Equation: _____
        ii. Correlation Coefficient: _____
    b. Logarithmic
        i. Equation: _____
        ii. Correlation Coefficient: _____
    c. Power
        i. Equation: _____
        ii. Correlation Coefficient: _____
    d. Which equation has a better correlation? _____

    Explain. _____

There are three methods to evaluate an unrounded regression equation at a specific value of *x*. The first two can also be used to find "*x*" at a given value of "*y*".

    1. Paste Y_1 at the end of the regression calculation before pressing ENTER by

        a. Pressing VARS.
        b. Go over to Y-VARS.

c. Choose 1:Function.
d. Choose 1:Y_1

e. Press ENTER. It will not appear that the calculator has calculated the regression differently but if Y= is pressed, the equation will be entered in Y_1.

f. The value can now be found on the GRAPH or in the TABLE.

2. After the regression has been calculated
   a. Press Y=.
   b. Press VARS.
   c. Choose 5:Statistics.
   d. Go over to EQ.
   e. Choose 1:RegEQ. The equation with unrounded coefficients should now appear in Y_1 and as above, the value can be found on the GRAPH or in the TABLE.

3. To evaluate the last regression calculated at a given value from the home screen:
   a. Store the desired value to x.
   b. Press VARS.
   c. Choose 5:Statistics.
   d. Go over to EQ.
   e. Choose 1:RegEQ.
   f. Press ENTER on the home screen.

Note:

A. To _____ is to find a specific value within the range of the given data.

B. To _____ is to find a specific value outside the range of the given data by extending the pattern found using the data.

Examples:

1. Using the "best equation", find the number of candles required to lift a balloon that has a volume of 18 cubic feet. _____

2. Find the number of candles required to lift a balloon that has a volume of 70 cubic feet. _____

3. Using the graph or table with the "best equation", find the number of candles required to lift 15.0 ounces. _____

Exponential Regressions

Graph $y = 2^x$ on the graphing calculator. Sketch the screen below.

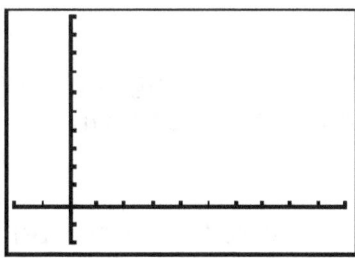

Describe the general form of an exponential equation:

Given a set of data points, the TI-83+/TI-84+ can find the exponential equation that best fits those points.

Use the table of world population below to answer the questions that follow.

World Population (Estimated)

Year:	1650	1750	1850	1900	1950	1980	2000	2001
Continent or Region:								
North America	5,000,000	5,000,000	39,000,000	106,000,000	221,000,000	372,000,000	481,000,000	486,000,000
South America	8,000,000	7,000,000	20,000,000	38,000,000	111,000,000	242,000,000	347,000,000	351,000,000
Europe	100,000,000	140,000,000	265,000,000	400,000,000	392,000,000	484,000,000	729,000,000	729,000,000
Asia	335,000,000	476,000,000	754,000,000	932,000,000	1,411,000,000	2,601,000,000	3,688,000,000	3,737,000,000
Africa	100,000,000	95,000,000	95,000,000	118,000,000	229,000,000	470,000,000	805,000,000	823,000,000
Oceania	2,000,000	2,000,000	2,000,000	6,000,000	12,000,000	23,000,000	31,000,000	31,000,000
Antarctica								
World	550,000,000	725,000,000	1,175,000,000	1,600,000,000	2,556,000,000	4,458,000,000	6,080,000,000	6,157,000,000

Source: 2002 World Almanac

```
ExPReg
```

Enter the years in L_1.

For each continent or region:

1. Enter the data in L_2.

2. Press STAT and choose CALC.

3. Choose 0:ExpReg.

4. On the home screen press VARS, choose Y-VARS.

5. Choose 1:Function

6. Choose 1:Y$_1$ then ENTER

7. Record the equation with coefficients rounded to the nearest hundredth.

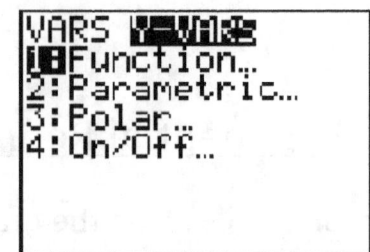

8. Before graphing the line, turning on a scatter plot of the data will make finding an appropriate window much easier. Press 2nd, Y= (STATPLOT).

9. Turn on Plot 1 and be sure that a scatter plot is chosen as the type, the x-list is L$_1$ and the y-list is L$_2$. If your calculator has been reset these should not need to be changed. This is the default setting for the stat plots.

10. Press ZOOM, choose 9:ZoomStat

11. The calculator should go immediately to the graphing screen and graph both the scatter plot and the regression equation. Carefully sketch the screen in the box provided. Include window settings.

12. Discuss how well the model fits the data.

13. Complete the table on the last page for each region before going on to the next region.

A. North America:

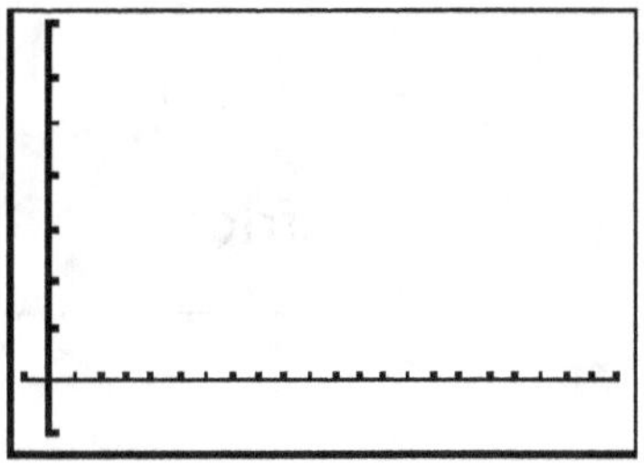

B. South America:

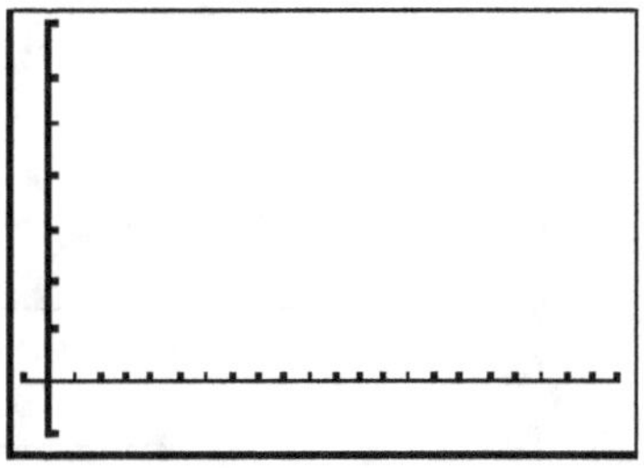

C. Europe:

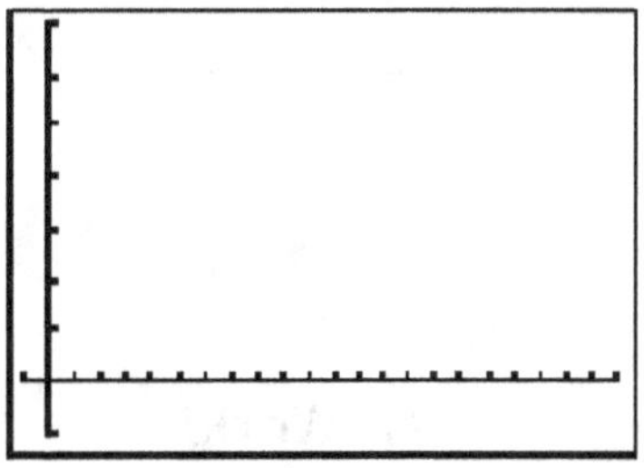

D. Asia:

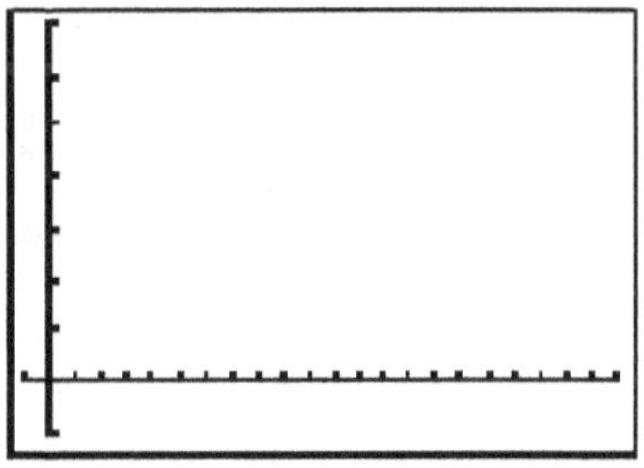

E. Africa:

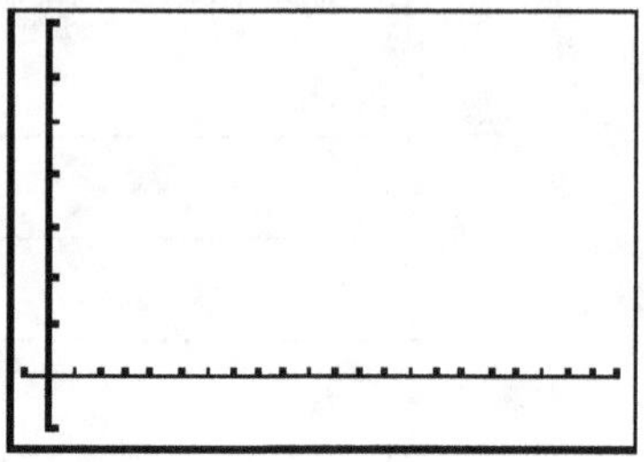

F. Oceania:

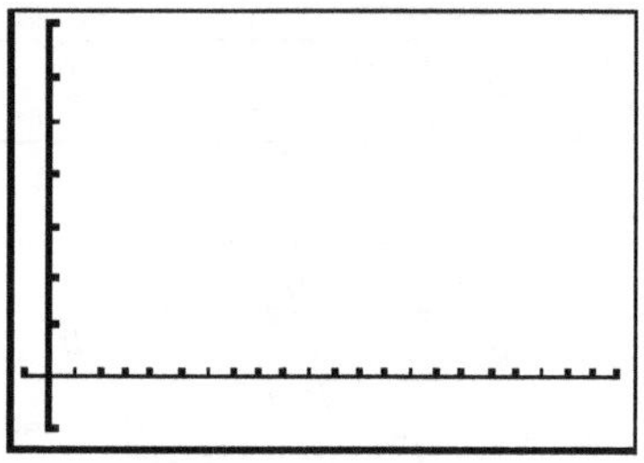

G. World:

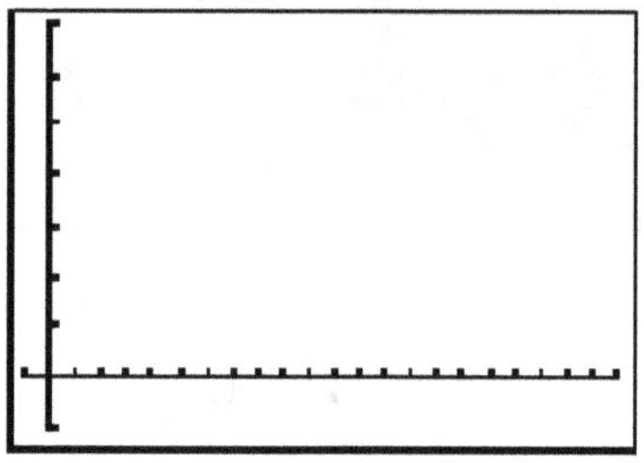

Based on the equations you found, complete the table below:

(Find the year given in the X column of the table and record the corresponding Y value. Recall that 2nd WINDOW (Table Setup) will allow you to go immediately to the required x value by entering it in TblStart and returning to the table.)

Population Estimates:

Year:	2010	2050	2103
Region:			
North America			
South America			
Europe			
Asia			
Africa			
Oceania			
World			

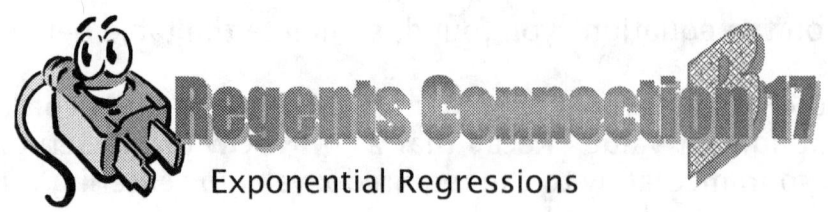

Exponential Regressions

<u>Aug '02, #32:</u> The breaking strength, *y*, in tons of steel cable with diameter *d*, in inches, is given in the table below.

d (in)	0.50	0.75	1.00	1.25	1.50	1.75
y (tons)	9.85	21.80	38.30	59.20	84.40	114.00

On the accompanying grid, make a scatter plot of these data. Write the exponential regression equation, expressing the regression coefficients to the nearest tenth.

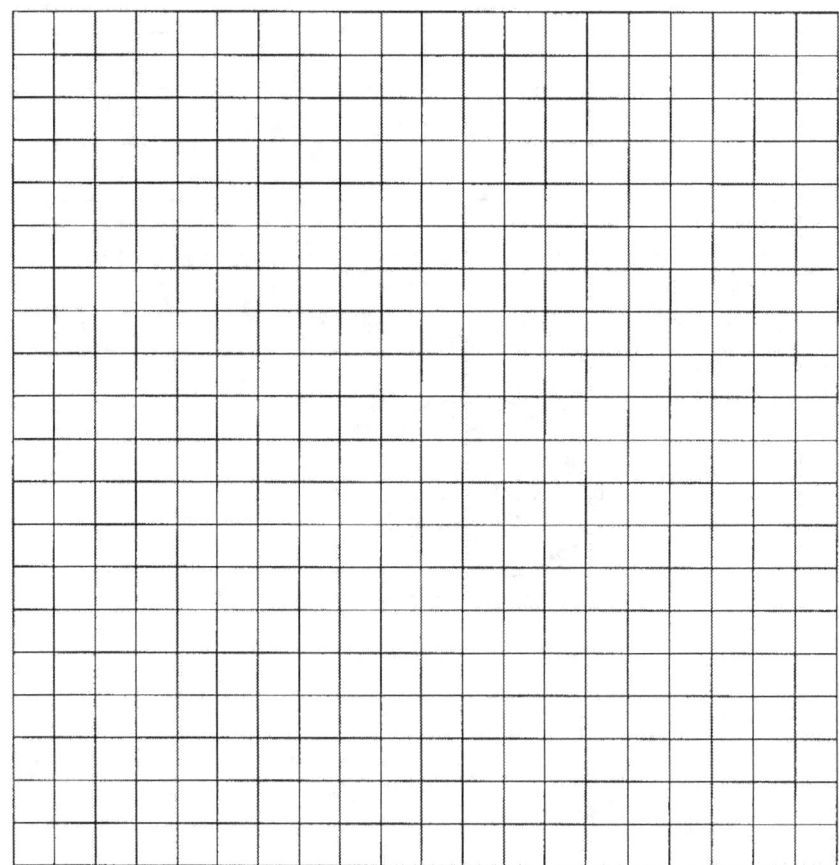

June '02, #34: The table below, created in 1996, shows a history of transit fares from 1955 to 1995. On the accompanying grid, construct a scatter plot where the independent variable is years. State the exponential regression equation with the coefficient and base rounded to the nearest thousandth. Using this equation, determine the prediction that should have been made for the year 1998, to the nearest cent.

Year	55.00	60.00	65.00	70.00	75.00	80.00	85.00	90.00	95.00
Fare ($)	0.10	0.15	0.20	0.30	0.40	0.60	0.80	1.15	1.50

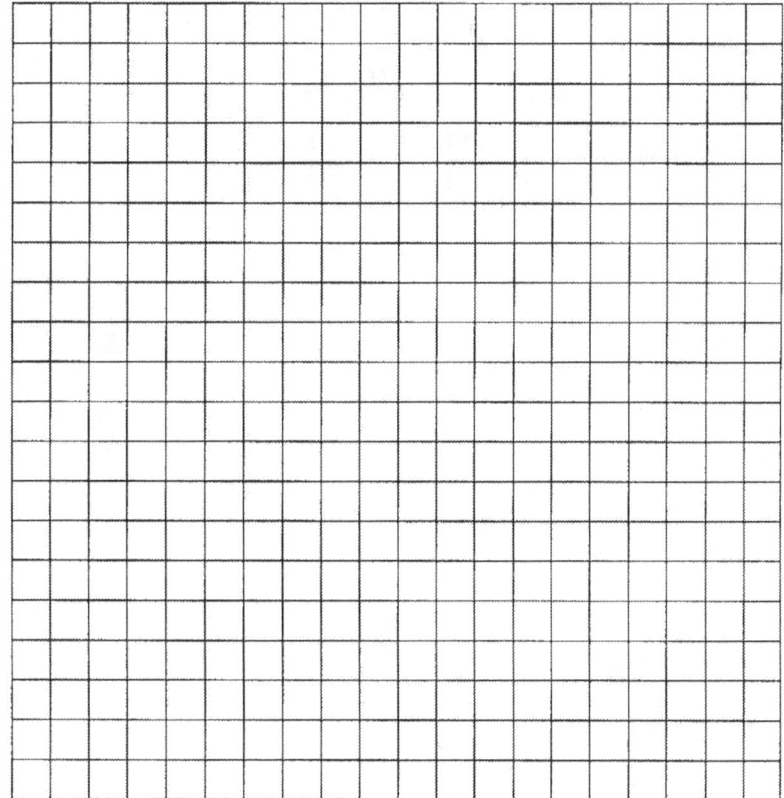

(Note: When asked to round coefficients in one part of a regents question, DO NOT use the unrounded regression equation to evaluate a given x or y value!!)

Sample #34: The volume of a particular gas was determined at various pressures. *P* is the pressure (in atmospheres) and is the independent variable on the horizontal axis, and *V* is the volume (in liters) and is the dependent variable on the vertical axis:
Create a scatter plot and find the equation of the curve of best fit. (Round answer constants to nearest tenth.)
Then, using the regression equation found, estimate *V* if *P*=2.5.

P	V
0.1	225
0.3	74.999
0.5	45
0.7	32.139
0.9	25
1.1	20.45
1.5	15
1.7	13.24
1.9	11.84
2.1	10.71
2.3	9.78

<u>Jan '04, #33:</u> The accompanying table shows the average salary of baseball players since 1984. Using the data in the table, create a scatter plot on the grid on the next page and state the exponential regression equation with the coefficient and base rounded to the nearest hundredth.

Using your written regression equation, estimate the salary of a baseball player in the year 2005, to the nearest thousand dollars.

Baseball Players' Salaries

Numbers of Years Since 1984	Average Salary (thousands of dollars)
0	290
1	320
2	400
3	495
4	600
5	700
6	820
7	1,000
8	1,250
9	1,580

Exponential & Logarithmic equations

Solving:

It is possible to solve exponential and logarithmic equations in SOLVER or by entering the left side of the equation in Y_1 and the right side of the equation in Y_2 and finding the point of intersection.

CAUTION!!
Because of the nature of the equations, it is possible to find a "false" solution!

Example:

Solve for x: $16^{x+4} = 32^{2x-10}$

```
16^(X+4)-32^(...=0
■X= -9.999796162...
 bound={-1E99,1...
■left-rt=0
```

The screen at the right makes it appear that the solution is approximately –10, but notice the three dots at the end of the number! Scrolling to the left as shown in the next screen, shows that the number is more accurately an extremely low negative number.

```
16^(X+4)-32^(...=0
■X=...61626373E98
 bound={-1E99,1...
■left-rt=0
```

This solution was found with an initial x-value of +10.5. Entering an initial value of 20 results in the "true" solution: $x=11$.

Graphing this problem is also difficult. When $x=11$, what is the value of 16^{x+4}? It is almost impossible to find a "good" window to show this point of intersection without a lot of extra work.

Similar problems arise with logarithmic equations.

Exploring exponential graphs:

Graph the exponential equations below and sketch in the space provided. Window: Zoom Standard then Zoom Decimal

1. $y = 3^x$

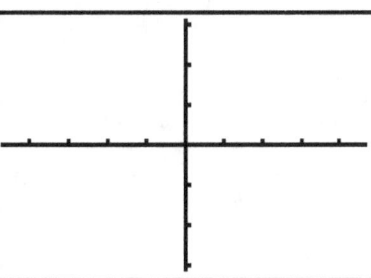

2. $y = 5^x$

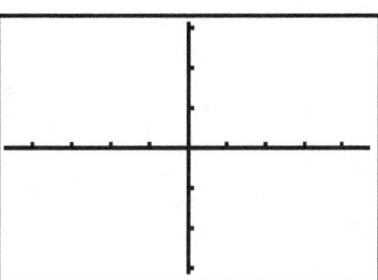

3. $y = 10^x$

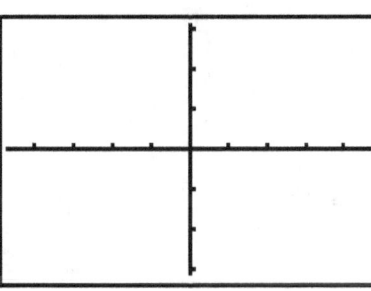

4. $y = .5^x$

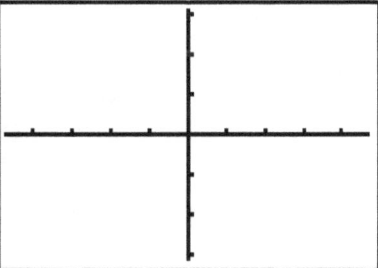

5. $y = .25^x$

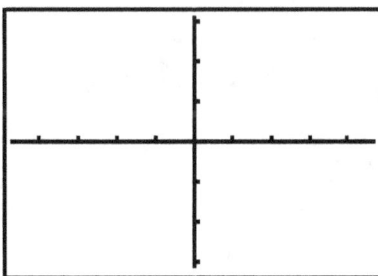

6. $y = -2^x$

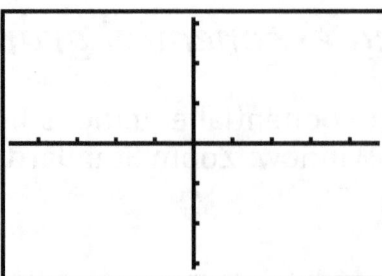

7. $y = -5^x$

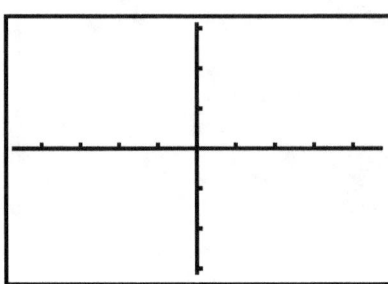

8. $y = 2^{2x}$

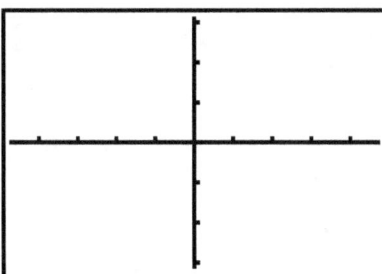

9. $y = 2^{5x}$

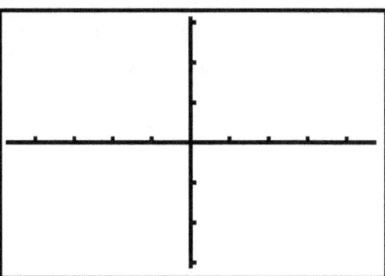

10. $y = 2^{.5x}$

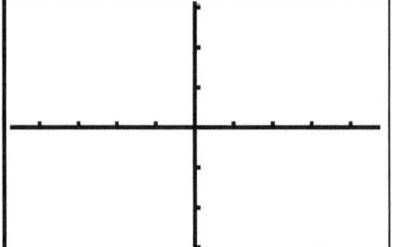

What effect does increasing the base number have?

What difference does having a base number between 0 and 1 make in the appearance of the graph?

How do negative base numbers change the appearance of the graph?

What point do all the graphs with positive bases have in common?

What would all graphs with negative bases have in common? _____

What effect does the coefficient on x have:

    when $x > 1$?

----

    when $0 < x < 1$

Logarithms:

There are three types of logarithms:

1. Base 10

2. Integer bases other than 10

3. Base e

The LOG key will calculate the "common" or base 10 logarithm. 2ⁿᵈ LOG will calculate 10^x or the antilogarithm.

To calculate logarithms with bases other than 10:

We'll do it the long way once.

Evaluate $\log_3 5$.

   a. Call the answer x so that we have

$$\log_3 5 = x$$

   b. Convert to exponential form:

$$3^x = 5$$

   c. Take the common log of both sides:

$$\log 3^x = \log 5$$

   d. Use the power rule for logarithms:

$$x \log 3 = \log 5$$

   e. Divide both sides by $\log 3$:

$$x = \frac{\log 5}{\log 3}$$

f. Substituting we have:

$$\log_3 5 = \frac{\log 5}{\log 3}$$

There was nothing special in the choice of the numbers used so we

can use the   rule $\boxed{\log_b a = \frac{\log a}{\log b}}$ to calculate logarithms with

bases other than 10.

Practice:

Evaluate each logarithm. Round decimals to the nearest thousandth.

    1. $\log_2 9$ _____

    2. $\log_5 12$ _____

    3. $\log_3 8$ _____

    4. $\log_4 .25$ _____

    5. Graph $y = \log_2 x$

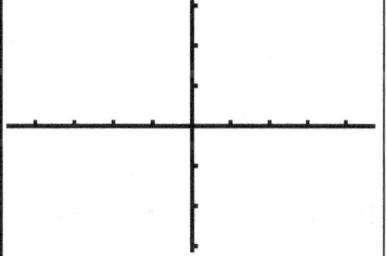

    6. Graph $y = \log_6 x$

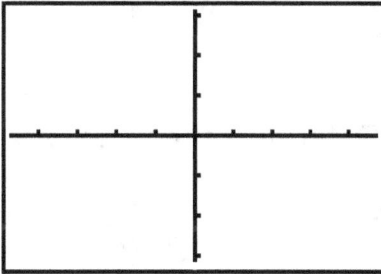

Exponential and Logarithmic
Functions

<u>Jan '03, #16:</u> The expression $\log 10^{(x+2)} - \log 10^x$ is equivalent to

    (1) –2   (3) 100
    (2) 2   (4) 1/100

<u>Jan '03, #24:</u> The relationship between the relative size of an earthquake, S, and the measure of the earthquake on the Richter scale, R, is given by the equation $\log s = R$. If an earthquake measured 3.2 on the Richter scale, what was its relative size to the nearest hundredth?

<u>Jan '03, #32:</u> Given: $f(x) = x^2$ and $g(x) = 2^x$
    a. The inverse of g is a function, but the inverse of f is not a function. Explain why this statement is true.

    b. Find $g^{-1}(f(3))$ to the nearest tenth.

<u>Aug '01, #10:</u> If $\log 5 = a$, then $\log 250$ can be expressed as

    (1) $50a$ (2) $2a + 1$ (3) $10 + 2a$ (4) $25a$

<u>Aug '01, #18:</u> Determine the value of x and y if $2^y = 8^x$ and $3^y = 3^{x+4}$.

    (1) *x*=6, *y*=2   (3) *x*=2, *y*=6
    (2) *x*=-2, *y*=-6   (4) *x*=*y*

Aug '01, #32: The amount A, in milligrams, of a 10-milligram dose of a drug remaining in the body after t hours is given by the formula $A=10(0.8)^t$. Find to the nearest tenth of an hour, how long it takes for half of the drug dose to be left in the body.

Aug '02, #4: What is the domain of $f(x)=2^x$?

    (1) all integers   (3) $x \geq 0$
    (2) all real numbers   (4) $x \leq 0$

Aug '02, #9: In the equation $\log_x 4 + \log_x 9 = 2$, x is equal to

    (1) $\sqrt{13}$ (2) 6   (3) 6.5   (4) 18

Aug '02, #12: If log $k=c$ log v + log p, k equals

    (1) $v^c p$ (2) $(vp)^c$ (3) v^c+p (4) $cv+p$

Aug '02, #24: The Franklins inherited \$3,500, which they want to invest for their child's future college expenses. If they invest it at 8.25% with interest compounded monthly, determine the value of the account, in dollars, after 5 years. Use the formula $A=P(1+\frac{r}{n})^{nt}$, where $A=$ value of the investment after t years, $P=$ principal invested, $r=$ annual interest rate, and $n=$ number of times compounded per year.

June '01, #2: The magnitude R of an earthquake is related to its intensity (I) by $R=log\left(\dfrac{I}{T}\right)$ where T is the threshold below which the earthquake is not noticed. If the intensity is doubled, its magnitude can be represented by

(1) $2(\log I - \log T)$
(2) $\log I - \log T$
(3) $2\log I - \log T$
(4) $\log 2 + \log I - \log T$

June '01, #15: The inverse of a function is a logarithmic function in the form $y = \log_b x$. Which equation represents the original function?

(1) $y = b^x$ (3) $x = b^y$
(2) $y = bx$ (4) $by = x$

June '01, #25: The scientists in a laboratory company raise amebas to sell to schools for use in biology classes. They know that one ameba divides into two amebas every hour and that the formula $t = \log_2 N$ can be used to determine how long in hours, t, it takes to produce a certain number of amebas, N. Determine to the nearest tenth of an hour, now long it takes to produce 10,000 amebas if they start with one ameba.

Sample #4: The expression $\log_2(x - 4)$ is undefined for all values of x such that

(1) $x > 1$ (2) $x > 0$ (3) $x \le 4$ (4) $x \le 0$

Sample #7: Solve for x: $64^{x-2} = 256^{2x} 64^{x-2} = 256^{2x}$

(1) $\dfrac{-6}{11}$ (2) $\dfrac{-6}{5}$ (3) $\dfrac{-1}{5}$ (4) 0

Sample #16: the population of Henderson City was 3,381,000 in 1994, and is growing at an annual rate of 1.8%. If this growth rate continues, what will the approximate population of Henderson City be in the year 2000?

(1) 3,696,000   (2) 3,763,000   (3) 3,798,000   (4) 3,831,000

Sample #27: Sketch the graph of the functions $f(x) = 3^x$ and $g(x) = \log_3 x$. Considering the graphs, describe the relationship between $f(x)$ and $g(x)$. Specify the domain and the range of g.

Sample #30: In the equation $y = 0.5(1.21^x)$, y represents the number of snowboarders in millions and x represents the number of years since 1988. Find the year in which the number of snowboarders will be 10 million for the first time. (Only an algebraic solution will be accepted.)

June '02, #24: Growth of a certain strain of bacteria is modeled by the equation $G = A(2.7)^{0.584t}$, where

> G=final number of bacteria
> A=initial number of bacteria
> t=time (in hours)

In approximately how many hours will 4 bacteria first increase to 2500 bacteria? Round your answer to the nearest hour.

June '02, #30: Solve for x: $\log_4(x^2 + 3x) - \log_4(x + 5) = 1$

<u>Jan '02, #8:</u> Which expression is *not* equivalent to $\log_b 36$?

    (1) $6\log_b 2$ (3) $2\log_b 6$

    (2) $\log_b 9 + \log_b 4$ (4) $\log_b 72 - \log_b 2$

<u>Jan '02, #30:</u> Depreciation (the decline in cash value) on a car can be determined by the formula $V=C(1-r)^t$, where V is the value of the car after t years, C is the original cost, and r is the rate of depreciation. If a car's cost, when new, is $15,000, the rate of depreciation is 30%, and the value of the car now is $3,000, how old is the car to the nearest tenth of a year?

<u>June '03, #1:</u> For which value of x is $y = \log x$ undefined?

    (1) 0 (2) $\dfrac{1}{10}$ (3) π (4) 1.483

<u>June '03, #3:</u> What is the value of x in the equation $81^{x+2} = 27^{5x+4}$?

    (1) $-\dfrac{2}{11}$ (2) $-\dfrac{3}{2}$ (3) $\dfrac{4}{11}$ (4) $-\dfrac{4}{11}$

<u>June '03, #16:</u> If $\log a = 2$ and $\log b = 3$, what is the numerical value of $\log \dfrac{\sqrt{a}}{b^3}$?

    (1) 8 (2) -8 (3) 25 (4) -25

<u>Jan '04, #9:</u> If $\log x = a$, $\log y = b$, and $\log z = c$, then $\log \dfrac{x^2 y}{\sqrt{z}}$ is equivalent to

    (1) $42a + b + \dfrac{1}{2}c$ (3) $a^2 + b - \dfrac{1}{2}c$

    (2) $2ab - \dfrac{1}{2}c$ (4) $2a + b - \dfrac{1}{2}c$

<u>Jan '04, #12:</u> The expression $\log_3(8-x)$ is defined for all values of x such that

    (1) $x > 8$ (2) $x \geq 8$ (3) $x < 8$ (4) $x \leq 8$

<u>Jan '04, #29:</u> The equation for radioactive decay is $p = (0.5)^{\frac{t}{H}}$, where p is the part of a substance with half-life H remaining after a period of time t.

A given substance has a half-life of 6,000 years. After t years, one-fifth of the original sample remains radioactive. Find t, to the nearest thousand years.

<u>Jan'04, #20:</u> The cells of a particular organism increase logarithmically. If g represents cell growth and h represents time, in hours, which graph best represents the growth pattern of the cells of this organism?

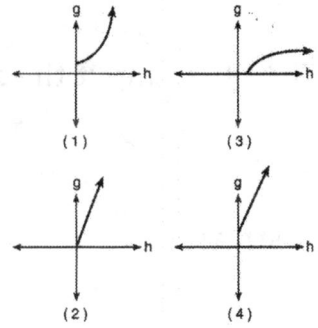

Days Between Dates

Did you ever want to quickly find out how many days were left until Christmas? Or how many days were between any two dates?

We will use an option on the Financial Application to find the number of days between any two dates.

This option is intended to be used to calculate loan payments but we can use it just for fun.

Press [APP] and choose 1. Finance.

Go to D:dbd(.

On the home screen input the first date by using MM.DDYY where M is the number of the month (a zero is not necessary if the month has only 1 digit), DD is the number of the day, and YY is last two digits of the year. ***Note that this will only work on years between 1950 and 2049!!***

Use a comma after the first date then input the second date in the same format.

Example:
How many days is it until Christmas?

Practice:
1. How many days is it until your graduation day?

----

2. How many days is it until the next Math A regents?

----

3. How many days is it until the next Math B regents?

_____

4. How many days are between your birthday and your best friend's birthday?

_____

5. How many days are there between Memorial Day and Labor Day this year?

_____

6. Choose three pairs of dates, state why they are significant, and give the number of days between these dates.

    a. _____

       _____

       _____

       _____

    b. _____

       _____

       _____

       _____

    c. _____

       _____

       _____

       _____

To use the finance application for compound interest questions:

1. Press APP.
2. Choose 1:Finance.
3. Choose 1:TVMSolver

The following variables will be listed:

Variable:	Used for:	Equivalent to:
N		
I%		
PV		
PMT		
P/Y		
C/Y		
FV		

Use ALHPA ENTER to solve as in EQUATION SOLVER.
(Don't change the rate to a decimal!)
Example:
A. If $5,000 is invested at 6.5% annual interest compounded monthly, how long will it take for the investment to double?

    N= _____   **Solve for N
    I%= _____
    PV= _____
    PMT= _____
    FV= _____
    P/Y= _____
    C/Y=_____

To convert to months:
1. Press Apps.
2. Choose 1: Finance.
3. Go over to VARS
4. Choose 1:N.
5. On home screen continue with *12.
6. _____

B. If $7500 is invested at 12% interest compounded yearly, how much is in the account after 5 years?

   N= _____
   I%= _____
   PV= _____
   PMT= _____
   FV= _____
   P/Y= _____
   C/Y=_____

C. What interest rate compounded annually is needed in order that a $16,000 investment grows to $50,000 in 18 years?

   N= _____
   I%= _____
   PV= _____
   PMT= _____
   FV= _____
   P/Y= _____
   C/Y=_____

**Note: changing P/Y changes C/Y so always change C/Y last if it is different than P/Y.

Continuous Compounding:

   Continuous compounding is done in the same manner but make C/Y very large, i.e. 1ᴇ11 is suggested. (1ᴇ4 is too small; 1ᴇ5 may round correctly but does not give an exact answer.)

Use the Finance Application to solve the following interest problems.

1. You deposit $1200 in an account that pays 4.5% annual interest compounded quarterly. What is the balance after 5 years?

   _____

2. You deposit $2250 in an account that pays 1.75% annual interest compounded continuously. How long will it take for the balance to double?

   _____

3. You deposit $500 in an account that pays 7.5% annual interest. Find the balance after 3 years if the interest is compounded
   a. Annually   _____

   b. Quarterly   _____

   c. Daily   _____

   d. Continuously   _____

4. Jason and Marcy each have $250. Jason plans to invest $50 for each of the next five years, while Marcy plans to invest all $250 now. Both accounts pay 4% annual interest compounded quarterly. Will they have the same amount of money after four years? If not, explain why.

   _____   _____

   _____

5. Is investing $3000 at 6% annual interest and $3000 at 8% annual interest equivalent to investing $6000 (the total of the two principals) at 7% annual interest (the average of the two interest rates)? Explain.

   _____   _____

   _____

6. You purchase an antique table for $500. Each year the value of the table increases by 4%. Complete the table of values for the first ten years.

Year	Value
1	
2	
3	
4	
5	
6	
7	
8	
9	
10	

7. You have inherited land that was purchased for $10,000 in 1960. The value of the land increased by approximately 6% per year. What will be the value of the land in 2010?

8. Can the Financial Application be used for *depreciation* problems? Use the following situation to find a method that will result in the correct answer. Check your answer using another method. Explain how you used the application so that someone unfamiliar with the calculator could answer any depreciation question with your method with no additional assistance.

> You buy a new car for $21,300. The value of the car decreases by 12.5% each year. Find the value of the car after 5 years.

Depreciation and Other Percent Problems

At the end of the last lesson you were asked to find a method for finding depreciation values. Any time a value is decreasing, entering a negative percent as the interest rate is appropriate.

Use the table below to find the value of each item after the given number of years.

Depreciation Rates:

Cars	18.75
Computers	37.50
Laptops	50.00
Lawn Mower	30.00
Refrigerator	7.50
Chain Saw	50.00
Truck	22.50

Source: http://www.financeselect.com.au/depcnrates.htm

1. If Josh buys a used car for $12,900 now, what will be the depreciated value of the car three years from now?

_____

2. If Rachel buys a truck for $18,700 now, what will be the depreciated value of the truck 5 years from now?

_____

3. Andrea is buying a computer for her home business. If its value now is $1500, how much depreciation can she claim on her income tax each of the next 5 years?

	Value	Change in Value
1st year:	_____	_____
2nd year:	_____	_____
3rd year:	_____	_____
4th year:	_____	_____
5th year:	_____	_____

4. Cathy has a lawn mowing business. She purchases a new lawn mower for $259.88. How much will the value of the lawn mower depreciate in 2 years?

_____

5. Beth is moving into an apartment while she attends college. She needs to by a new refrigerator. If the refrigerator is worth $420 now, what will its value be when Beth finishes college in 4 years?

_____

6. Mindy is going into the firewood business and purchases a chain saw for $575. How much will the chainsaw depreciate in value over the next three years?

1st year: _____
2nd year: _____
3rd year: _____

7. Kim is claiming her laptop computer as a business expense. If the laptop was $1450 new, what will she be claiming as depreciation in its value in the third year she owns it?

_____

Other percent problems:

Other percents can also be calculated in Finance. Simple percents should be entered as the interest rate with 1 as the number of years, the "is" will be the future value (FV) and the "of" will be the present value (PV). Leave other values at their default settings. Either the present value or future value will need to be negated when entered.

8. Jenn is purchasing an entertainment center for her apartment. It is on sale for $648 and she can defer payment for 2 years with a finance charge of 24% per year. What will the entertainment center cost Jenn if she waits the full two years to pay?

_____

9. Kaitlin is buying a new four-piece bedroom set for $1899. If sales tax is 7¼%, what is the total cost of the bedroom set?

_____

10.   A new gas grill is advertised on sale for $149.99. The ad claims this is a savings of $30. What percent is this savings of the original price?

_____

11.   A 2003 Mustang Convertible is advertised for $17988. If Alan pays $2000 down and pays the balance at 7.9% interest over 4 years, what will his monthly payment be? (Use N=48, future value =0, and payment per year=12.)

_____

12.   Alan (from #11) is offered $3000 for his old car he has decided to trade in.
   a. By how much will his monthly payment change?

_____

   b. If sales tax is 7.5%, how much extra money will Alan need at the time of the purchase?

_____

   c. What will be the depreciated value of the car when Alan has completed his payments?

_____

13.   A 2003 Ranger 4x4 is advertised for$13988.
   a. If Pat pays $2000 down with a $1500 trade-in, and 8.9% interest for 3 years, what will his monthly payment be?

_____

   b. What will the sales tax on the truck be at 7.25%?

_____

   c. What will the total cost of the truck be after tax and interest?

_____

   d. What will the depreciated value of the truck be after Pat has completed his payments?

_____

14.   Gary has decided to buy a house for $122,000.He has enough cash for a $5000 down payment and all closing costs.
   a. If he can get a mortgage for 3.8% interest compounded continually for 20 years, what will his monthly payment be?

_____

   b. What should Gary be prepared to pay for sales tax if the rate is 7.25%?

_____

Applications of Finance

<u>Jan '02, #30:</u> Depreciation (the decline in cash value) on a car can be determined by the formula $V = C(1-r)^t$, where V is the value of the car after t years, C is the original cost, and r is the rate of depreciation. If a car's cost, when new, is $15,000, the rate of depreciation is 30%, and the value of the car now is $3,000, how old is the car to the nearest tenth of a year?

<u>Sample #16:</u> The population of Henderson City was 3,381,000 in 1994, and is growing at an annual rate of 1.8%. If this growth rate continues, what will the approximate population of Henderson City be in the year 2000?

    (1) 3,696,000   (2) 3,763,000   (3) 3,798,000 (4) 3,831,000

<u>Aug '02, #24:</u> The Franklins inherited $3,500, which they want to invest for their child's future college expenses. If they invest it at 8.25% with interest compounded monthly, determine the value of the account, in dollars, after 5 years. Use the formula $A = P\left(1+\dfrac{r}{n}\right)^{nt}$, where

A= value of the investment after t years, P= principal invested, r= annual interest rate, and n= number of times compounded per year.

<u>June '03, #30:</u> Sean invests $10,000 at an annual rate of 5% compounded continuously, according to the formula $A=Pe^{rt}$, where A is the amount, P is the principal, $e=2.718$, r is the rate of interest, and t is time, in years.
Determine, to the nearest dollar, the amount of money he will have after 2 years.
Determine how many years, to the nearest year, it will take for his initial investment to double.

Transformations With Lists!

A rule for a transformation can be applied to lists to quickly produce a graph of the transformation.

We need to start with a polygon in the plane. Suppose we are looking at the triangle MAY with vertices M(-1,1), A(7,-2), and Y(1,7).

1. Enter the *x*'s in list 1 and the *y*'s in list 2 in the order they are listed.
2. Repeat the first coordinates at the bottom of the list.
3. Turn on Stat Plot 1.
4. Choose a line graph.
5. Graph.

For consistency in this lesson, press Zoom Standard then Zoom Square.

Sketch the final screen below.

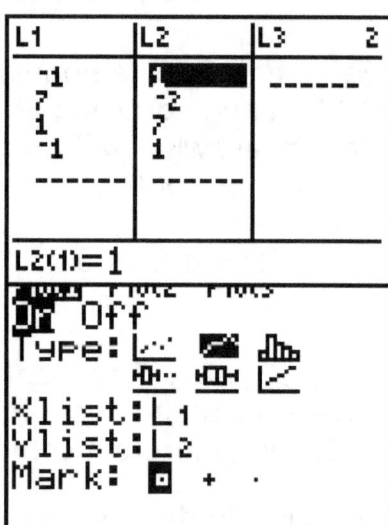

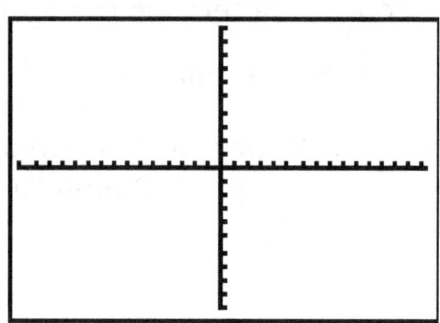

The "rules" for transformations in the plane are given on the next page. We will apply the rule to the *x* and *y* lists and adjust Stat Plot 2 accordingly.

Summary of Coordinate "Rules" for Transformations:

Line Reflections	
Reflection in the x-axis:	$r_{x-axis}(x,y)=(x,-y)$
Reflection in the y-axis:	$r_{y-axis}(x,y)=(-x,y)$
Reflection in the line y=x:	$r_{y=x}(x,y)=(y,x)$
Reflection in the line y=-x:	$r_{y=-x}(x,y)=(-y,-x)$
Point Reflection in the Origin:	$R_0(x,y)=(-x,-y)$
Rotations about the Origin:	
+90° or -270°:	$R_{90°}(x,y)=(-y,x)$
+/-180°:	$R_{180°}(x,y)=(-x,-y)$
+270° or -90°:	$R_{270°}(x,y)=(y,-x)$
Translation:	$T_{a,b}(x,y)=(x+a,y+b)$
Dilation:	$D_k(x,y)=(kx,ky)$

Example:

A. Sketch the image of triangle MAY after the transformation r_{x-axis}.

1. Note that the "rule" for a reflection in the y-axis leaves y unchanged but negates x. Create List 3 by entering –L$_1$ as a formula in L$_3$.

L1	L2	⬛3	3
-1	1	------	
7	-2		
1	7		
-1	1		
------	------		

L3 = -L1

2. Set up Stat Plot 2 as a line graph of L$_3$ and L$_2$.

3. Graph and sketch the screen below.

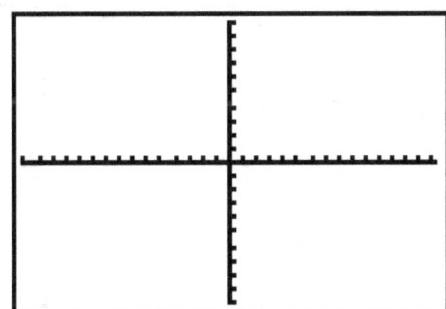

Practice:

Create line plots that display the original figure and its image under the given transformation on the same screen.

1. C(2,7), A(-3,6), T(-5,-3); $T_{3,4}$
   ($L_3=L_1+3$, $L_4=L_2+4$, use L_3 and L_4 in Stat Plot 2)

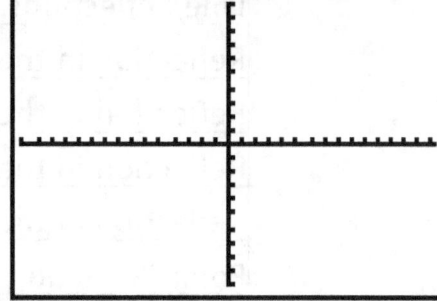

2. M(1,0), A(-2,-5), T(-4,3), H(2,5); $R_{180°}$

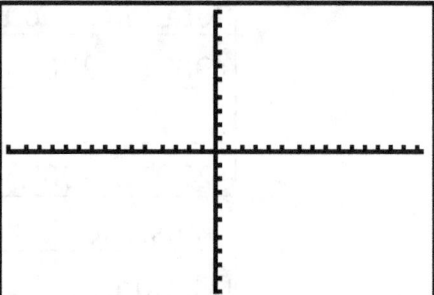

3. P(-3,2), Q(-6,6), R(-6,1), S(-3,6), T(-8,4); $r_{y\text{-axis}}$

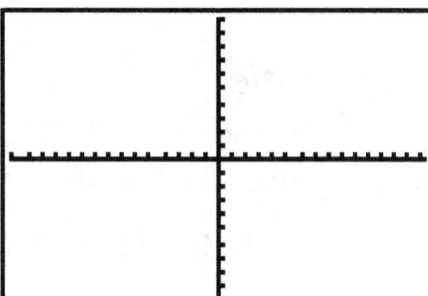

4. A(3,3), B(7,2), C(6,4), D(6,7); $R_{270°}$

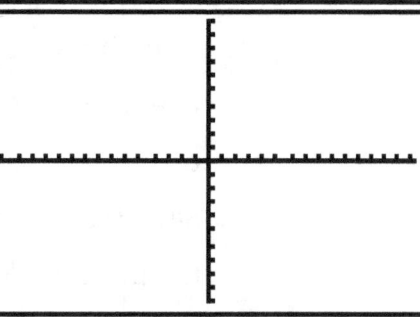

5. U(2,2), V(5,2), W(6,4), X(5,6), Y(2,6), Z(3,4); $r_{x\text{-axis}}$

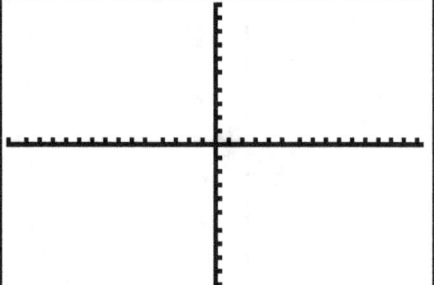

6. B(-1,0), U(-1,2), T(-3,3),   E(-5,1),
   R(-4,-1), F(-1,0), L(-5,-4), Y(-5,-2);
   $r_{y=x}$

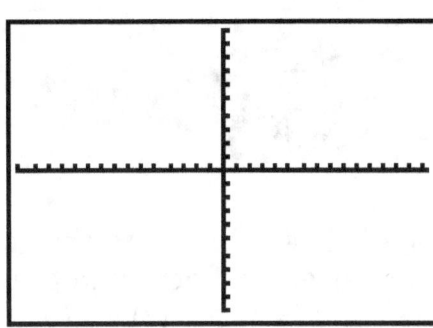

7. F(-2,2), L(-2,0), O(-3,-1),W(-1,-1),
   E(-1,-3), R(0,-2), S(2,-2); $r_{y=-x}$

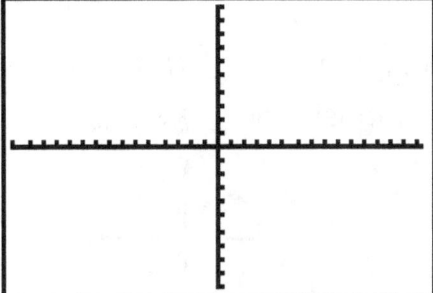

8. T(0,0), R(2,-5), Y(6,-2); $R_{90°}$

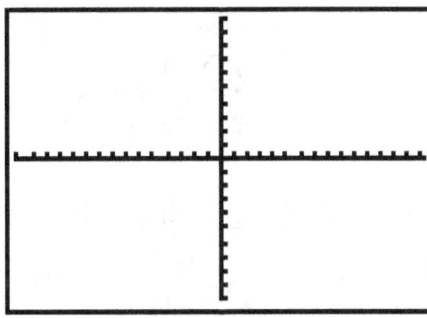

9. T(1,1), R(-1,1), A(-3,-2), P(3,-2); D_3

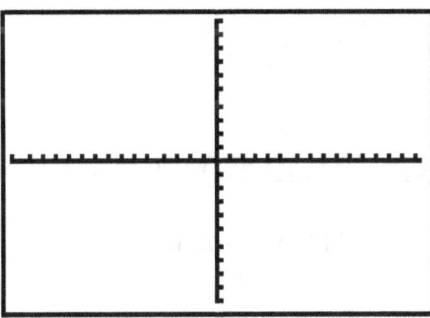

Challenge:
   10.   S(1,-1), T(0,2), A(-1,-1),
         R(1,1), Y(-1,1); $r_{y-axis} \circ D_5 \circ R_{270°}$

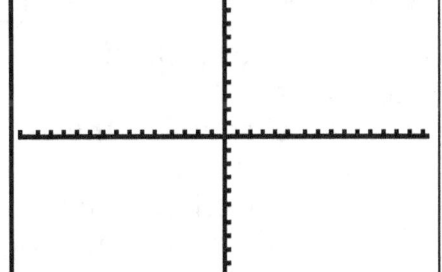

Transformations

June '03, #6: What are the coordinates of point P, the image of point (3,-4) after a reflection in the line *y=x*?
    (1) (3,4)   (2) (-3,4)   (3) (4,-3)   (4) (-4,3)

June '03, #9: If $f(x) = \cos x$, which graph represents $f(x)$ under the composition $r_{y-axis} \circ r_{x-axis}$?

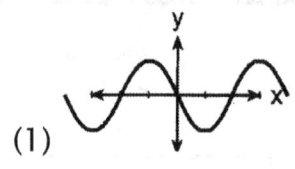

(1)

(3)

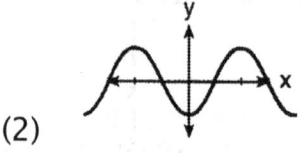

(2)

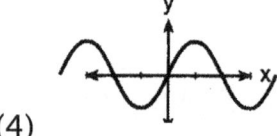

(4)

June '02, #17: Point P' is the image of point P(-3,4) after a translation defined by $T_{(7,-1)}$. Which other transformation on P would also produce P'?

    (1) $r_{y=-x}$ (3) $R_{90°}$

    (2) r_{y-axis} (4) $R_{-90°}$

June '02, #18: Which transformation does not preserve orientation?

    (1) translation   (3) reflection in the *y*-axis
    (2) dilation   (4) rotation

Sample #8: If $y = 2^x$ and $y = \left(\dfrac{1}{2}\right)^x$ are graphed on the same set of coordinate axes, which transformation would map one of these curves onto the other?
    (1) reflection in the *y*-axis
    (2) reflection in the *x*-axis
    (3) reflection in the line *y=x*
    (4) reflection in the origin

<u>June '01, #29</u>: Two parabolic arches are to be built. The equation of the first arch can be expressed as $y = -x^2 + 9$, with a range of $0 \le y \le 9$, and the second arch is created by the transformation $T_{7,0}$. On the accompanying set of axes, graph the equations of the two arches. Graph the line of symmetry formed by the parabola and its transformation and label it with the proper equation.

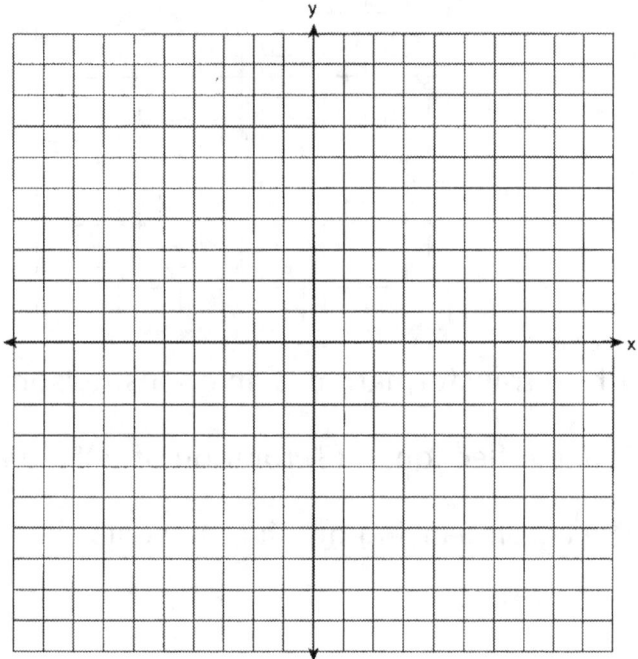

<u>Aug '01, #5</u>: Which transformation is a direct isometry?

(1) D_2 (3) $r_{y\text{-axis}}$

(2) D_{-2} (4) $T_{2,5}$

<u>Jan '02, #10</u>: Which transformation is not an isometry?

(1) $r_{y=x}$ (3) $T_{3,6}$

(2) $R_{0,90°}$ (4) D_2

Jan '02, #32: a. On the accompanying grid, graph the equation $2y = 2x^2 - 4$ in the interval $-3 \le x \le 3$ and label it "a".
b. On the same grid, sketch the image of a under $T_{5,-2} \circ r_{x\text{-axis}}$ and label it "b".

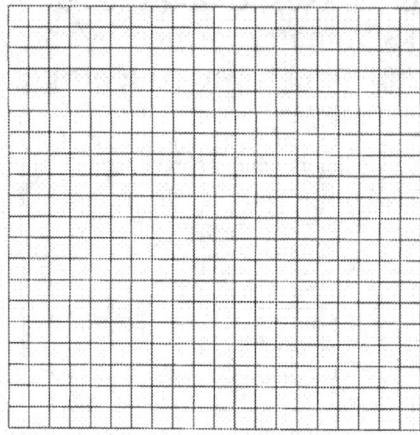

June '03, #13: Which transformation is an opposite isometry?

(1) dilation   (2) line reflection   (3) rotation of 90°   (4) translation

Aug '02, #19: The accompanying graph represents the figure I .

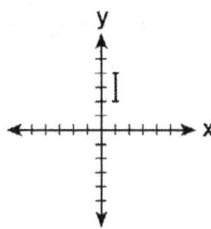

Which graph represents I after a transformation defined by $r_{y=x} \circ R_{90°}$?

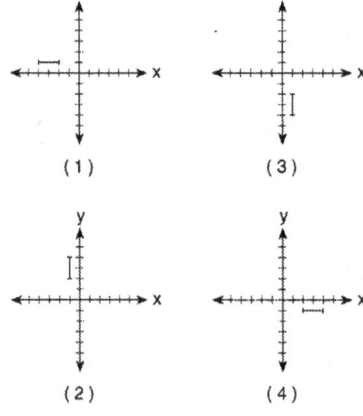

<u>Aug '02, #31:</u> Graph and label the following equations, a and b, on the accompanying set of coordinate axes.

a: $y = x^2$

b: $y = -(x-4)^2 + 3$

Describe the composition of transformations performed on a to get b.

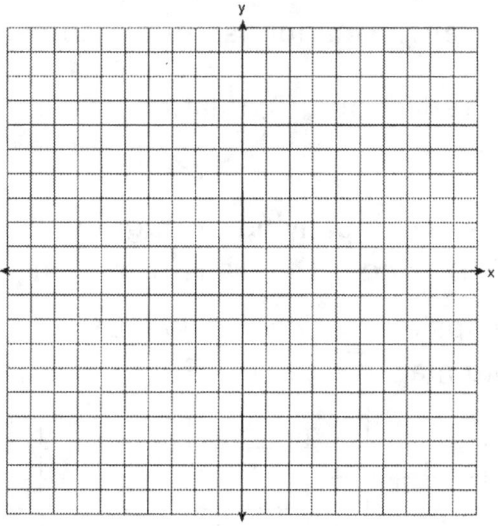

<u>Aug '01, #15:</u> The graph of $f(x)$ is shown in the accompanying diagram.

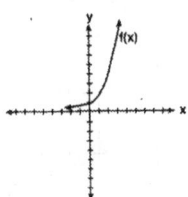

Which graph represents $f(x)r_{x-axis} \circ r_{y-axis}$?

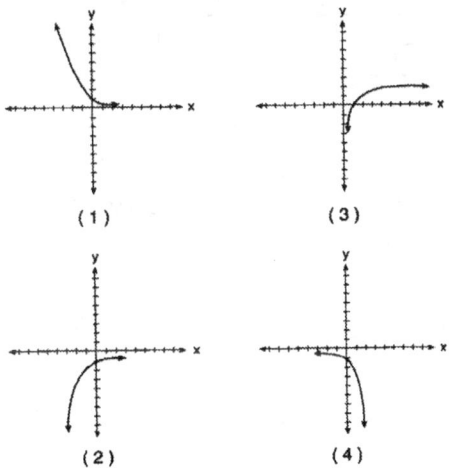

(1)   (3)

(2)   (4)

Jan '03, #11: In which quadrant would the image of point (5,-3) fall after a dilation using a factor of –3?

(1) I   (2) II   (3) III   (4) IV

Aug '03, #8: Which transformation is not an isometry?

(1) rotation   (2) line reflection   (3) dilation   (4) translation

Aug '03, #27: On the accompanying grid, graph and label $\overline{AB}$, where A is (0,5) and B is (2,0). Under the transformation $r_{x-axis} \circ r_{y-axis}\left(\overline{AB}\right)$, A maps to A", and B maps to B". Graph and label $\overline{A"B"}$. What single transformation would map $\overline{AB}$ to $\overline{A"B"}$?

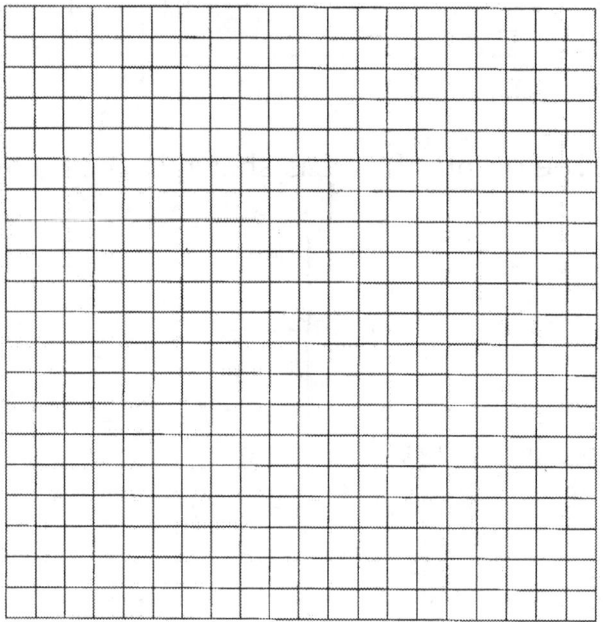

Jan '04, #19: What is the axis of symmetry of the graph of the equation $x = y^2$?

(1) x-axis   (2) y-axis   (3) line y=x   (4) line y=-x

Compositions and Inverses of Functions

While the TI-83+/TI-84+ will not find the equation of a composition of functions or the equation of a function's inverse directly, it can be used to find the numerical value of a composition or the graph of a composition or inverse, and in some cases, with a little more work, the equation.

Compositions:
Remember to work from the inside out when composing functions!

Suppose you need to know what $g(f(3))$ is when $g(x) = x^2$ and $f(x) = 4x - 3$.

1. Enter the inner function in Y_1.

2. Enter the outer function in Y_2 but use Y_1 as the variable instead of x. To use Y_1 as the variable you will need to get Y_1 from the variable menu:

   VARS   →   Y-VARS, choose 1:Function, then 1:Y_1.

3. If you only need to see the graph, turn off Y_1 by moving the cursor to the equal sign and pressing ENTER. Press GRAPH .

4. If you need the value of the composition at a particular value of x, press 2nd   GRAPH then look up that value of x in the table, using the value in the Y_2 column as your answer.

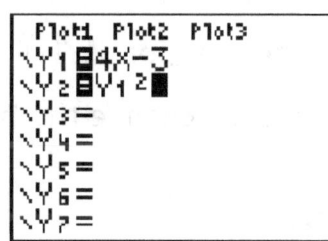

Example:

Let $f(x) = x - 5$ $\qquad$ $g(x) = 2x7$ $\qquad$ $h(x) = x^2 - 3$

1. Find $g(f(3))$ $\quad$ _____ $\qquad$ 2. Find $f(g(-2))$ $\quad$ _____

3. Find $h(g(0))$ $\quad$ _____ $\qquad$ 4. Find $g(h(5))$ $\quad$ _____

5. Find $h(f(6))$ $\quad$ _____ $\qquad$ 6. Find $f(g(h(1)))$ _____

If you need the equation of the composition you could work out the algebra yourself, then you can check your answer on the calculator.

Example:

Let $f(x) = x^2 + 3$ $\qquad$ $g(x) = 3x - 1$

Find $g(f(x))$ the "long" way:

Enter $f(x)$ in Y_1, $g(f(x))$ in Y_2 as we did earlier. Now enter your equation for $g(f(x))$ in Y_3. To better see the distinction between Y_2 and Y_3, change the type of line in front of Y_3 to the circle with a "tail".

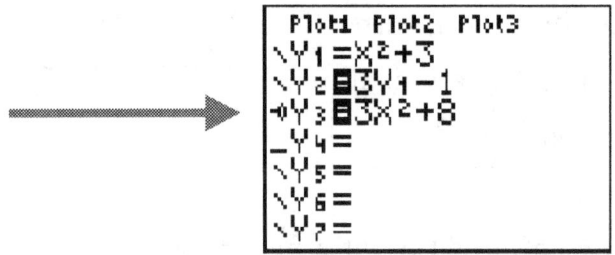

Now graph and watch to see if the circle traces the Y_2 "line".

If your graph is too "busy", Y_1 can be turned off by placing the cursor on the equal sign after Y_1 and pressing ENTER.

OR:

We could use a regression to find the equation:

a. Enter the composition as before. Be sure to turn Y_1 off. (See step 3 on the previous page.) The purpose in turning Y_1 off is to be able to view only the composition.

b. From the TABLE, choose the coordinates of three points and list these on your paper for use in a later step.

c. Check the graph to see what type of function the composition is (linear or quadratic).

d. Enter the x's from the coordinates you wrote down in L_1 and the corresponding y's in L_2.

e. Press STAT.

f. Go over to CALC.

g. If the graph was linear choose 4:LinReg. If the graph was quadratic (a parabola) choose 5:QuadRed. Press ENTER. Press ENTER again on the home screen (unless you needed to list your coordinates in lists other than L_1 and L_2, then you will need to input the list names with a comma between them then press ENTER.)

Example:

Let $g(x) = 4x^2$ and $f(x) = 2x - 8$. Find then equation for $f(g(x))$.

1. Enter equations in Y_1 and Y_2 and turn Y_1 off. $\longrightarrow$

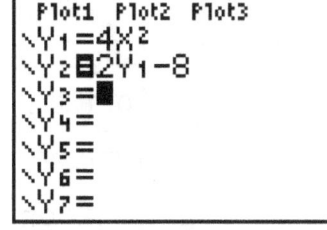

2. Check the graph to see which type of regression you will need to use. This one will be a _____ equation.

3. Find three points on the composition graph from the TABLE.

    a. _____

    b. _____

    c. _____

4. Use these three points to do a quadratic regression:

_____

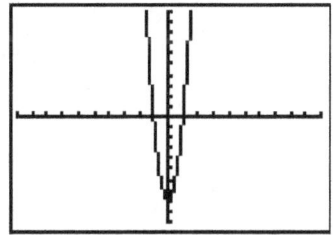

Practice:

Find the equation for each composition below. Also list the three points used and the type of regression required.

1. $f(x) = 3x^2$, $g(x) = 2x - 5$

    a. Points: _____ _____ _____

    b. Type of regression: _____

    c. Equation of composition: $f(g(x)) =$ _____

2. $g(x) = 3x - 7$, $h(x) = 6x - 3$

    a. Points: _____ _____ _____

    b. Type of regression: _____

    c. Equation of composition: $g(h(x)) =$ _____

3. $f(x) = x^2 - 1$, $g(x) = 5x$

    a. Points: _____ _____ _____

    b. Type of regression: _____

    c. Equation of composition: $g(f(x)) =$ _____

4. $f(x) = 2x + 6$, $h(x) = -2x + 6$

    a. Points: _____ _____ _____

    b. Type of regression: _____

    c. Equation of composition: $f(h(x))$ _____

5. $f(x) = 3x^2 - 5x + 10$, $g(x) = 3x - 4$

    a. Points: _____ _____ _____

    b. Type of regression: _____

    c. Equation of composition: $f(g(x)) =$ _____

6. **Challenge: $f(x) = 4x^2 + 3x - 5$, $g(x) = 3x - 1$, $h(x) = 7x + 2$

    a. Points: _____ _____ _____

    b. Type of regression: _____

    c. Equation of composition: $f(g(h(x))) =$ _____

Inverses:

Inverses can't be directly graphed but they can be "drawn".

When you have drawn the inverse, you can check an answer as above.

Consider the function $f(x) = 6x - 5$.

1. Enter this function in Y_1.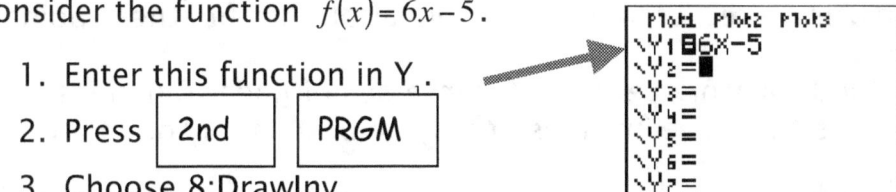

2. Press [2nd]   [PRGM]

3. Choose 8:DrawInv.

4. Paste Y_1 after DrawInv when it appears on the homescreen.

5. Press [ENTER] .

6. You should see the graphing screen now. Note that because the inverse is only drawn and not graphed, the TRACE and CALCULATE choices will not work on this system of equations.

7. If you have found the equation for the inverse, you can check your answer by entering it in Y_2 and following the directions for checking compositions. ***Note that the order in which graphing and drawing are done will sometimes erase the drawing.

Practice:

A. If $f(x) = 4x - 3$, find and check $f^{-1}(x)$.

B. If $f(x) = x - 12$, find and check $f^{-1}(x)$.

As with the compositions, there is a regression method to find the equation of the inverse of a function:

The inverse of a function can be viewed as a _____ in the line _____ .

Recall that the "rule" for this reflection can be written as

$$r_{y=x}(a,b) = (b,a)$$

In other words, the x and y values "switch places". The general idea will be to identify some x,y pairs, switch them, then using a regression, find an equation that fits these points.

***This method will work very well for inverses of <u>linear</u> functions. For other functions that have inverses, it may be best to rely on other methods.

Suppose you need to find the inverse of the following function:

$$y = \frac{2}{3}x - 12$$

1. Enter the original function in Y_1.

2. Go to the TABLE and choose 3 x,y pairs:
   a. _____

   b. _____

   c. _____

3. Enter the y's in L_1 and the x's in L_2.

4. Press STAT, go over to CALC, then choose 4:LinReg.
5. Press ENTER.
6. Write the regression equation: _____
7. To check:
   a. Go to Y=.
   b. In Y_2, enter the equation as you wrote it, or

   c. Have the calculator enter it by pressing Vars.
   d. Choose 5:Statistics.
   e. Go over to EQ.
   f. Choose 1:RegEQ.
   g. Press GRAPH.

h. Do Y_1 and Y_2 appear to be reflections of each other in the line

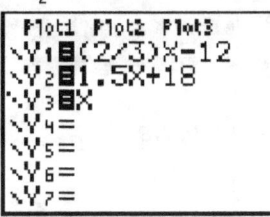 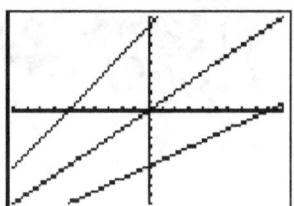

$y=x$?

Practice:

Name the points used for the linear regression then write the regression equation for the inverse of each function below.

1. $f(x) = 3x - 10$

    a. _____ _____ _____

    b. $f^{-1}(x) =$ _____

2. $f(x) = \dfrac{2}{5}x + 7$

    a. _____ _____ _____

    b. $f^{-1}(x) =$ _____

3. $f(x) = 4x + \dfrac{2}{7}$

    a. _____ _____ _____

    b. $f^{-1}(x) =$ _____

4. $f(x) = 5x + 14$

    a. _____ _____ _____

    b. $f^{-1}(x) =$ _____

5. $f(x) = \dfrac{1}{11}x + 8$

    a. _____ _____ _____

    b. $f^{-1}(x) =$ _____

6. $f(x) = 7 + \dfrac{1}{3}x$

    a. _____ _____ _____

    b. $f^{-1}(x) =$ _____

Functions

June '02, #16: Which diagram represents a one-to-one function?

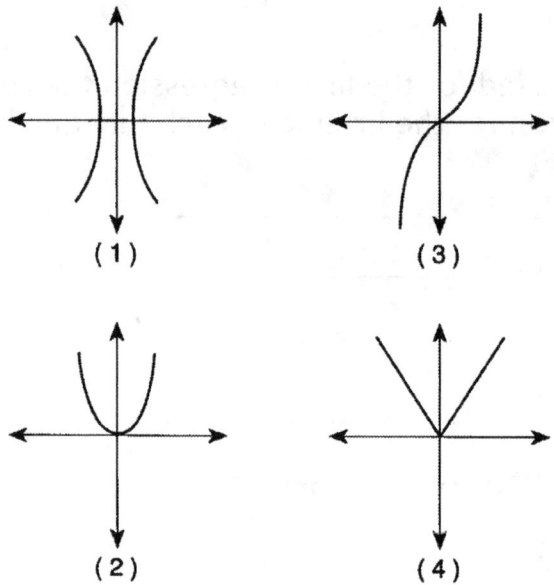

(1)

(3)

(2)

(4)

June '02, #20: Which graph represents the inverse of $f(x)$={(0,1), (1,4), (2,3)}?

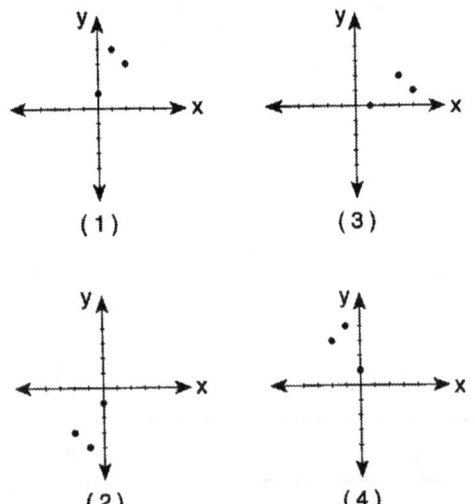

(1)

(3)

(2)

(4)

<u>Jan '02, #7:</u> If $f(x) = 5x^2$ and $g(x) = \sqrt{2x}$, what is the value of $(f \circ g)(8)$?

    (1) $8\sqrt{10}$ (2) 16   (3) 80   (4) 1,280

<u>Jan '02, #9:</u> If a function is defined by the equation $y = 3x + 2$, which equation defines the inverse of this function?

    (1) $x = \dfrac{1}{3}y + \dfrac{1}{2}$ (3) $y = \dfrac{1}{3}x - \dfrac{2}{3}$

    (2) $y = \dfrac{1}{3}x + \dfrac{1}{2}$ (4) $y = -3x - 2$

<u>Jan '02, #11:</u> Which relation is a function?

    (1) $x = 4$ (3) $y = \sin x$

    (2) $x = y^2 + 1$ (4) $x^2 + y^2 = 16$

<u>Aug '02, #5:</u> A function is defined by the equation $y = 5x - 5$. Which equation defines the inverse of this function?

    (1) $y = \dfrac{1}{5x - 5}$ (3) $x = \dfrac{1}{5y - 5}$

    (2) $y = 5x + 5$ (4) $x = 5y - 5$

<u>Aug '02, #16:</u> If point (a,b) lies on the graph $y = f(x)$, the graph $y = f^{-1}(x)$ must contain point

    (1) (b,a)   (2) (a,0)   (3) (0,b)   (4) (-a,-b)

<u>Aug '02, #4:</u> What is the domain of $f(x) = 2^x$?

    (1) all integers   (3) $x \geq 0$
    (2) all real numbers   (4) $x \leq 0$

<u>Aug '01, #1</u>: Which relation is not a function?

(1) $y = 2x + 4$ (3) $x = 3y - 2$

(2) $y = x^2 - 4x + 3$ (4) $x = y^2 - 6x + 8$

<u>June '02, #13</u>: Which equation represents a function?

(1) $4y^2 = 36 - 9x^2$ (3) $x^2 + y^2 = 4$

(2) $y = x^2 - 3x - 4$ (4) $x = y^2 - 6x + 8$

<u>Sample #3</u>: If the graph of $f(x)$ is which of the following is the graph of $-f(x)$?

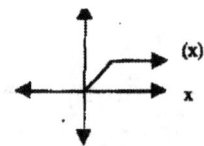

(1) 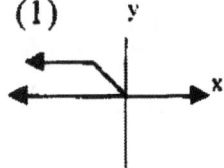   (2) 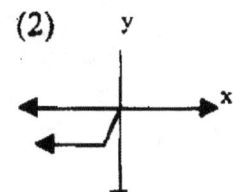   (3) 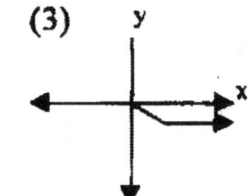   (4)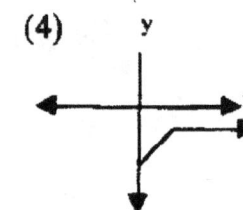

<u>June '02, #10:</u> If $f(x) = 2x^2 + 4$ and $g(x) = x - 3$, which number satisfies $f(x) = (f \circ g)(x)$?

(1) $\dfrac{3}{2}$ (2) $\dfrac{3}{4}$ (3) 5   (4) 4

<u>June '01, #30:</u> Draw $f(x) = 2x^2$ and $f^{-1}(x)$ in the interval $0 \leq x \leq 2$ on the accompanying set of axes. State the coordinates of the points of intersection.

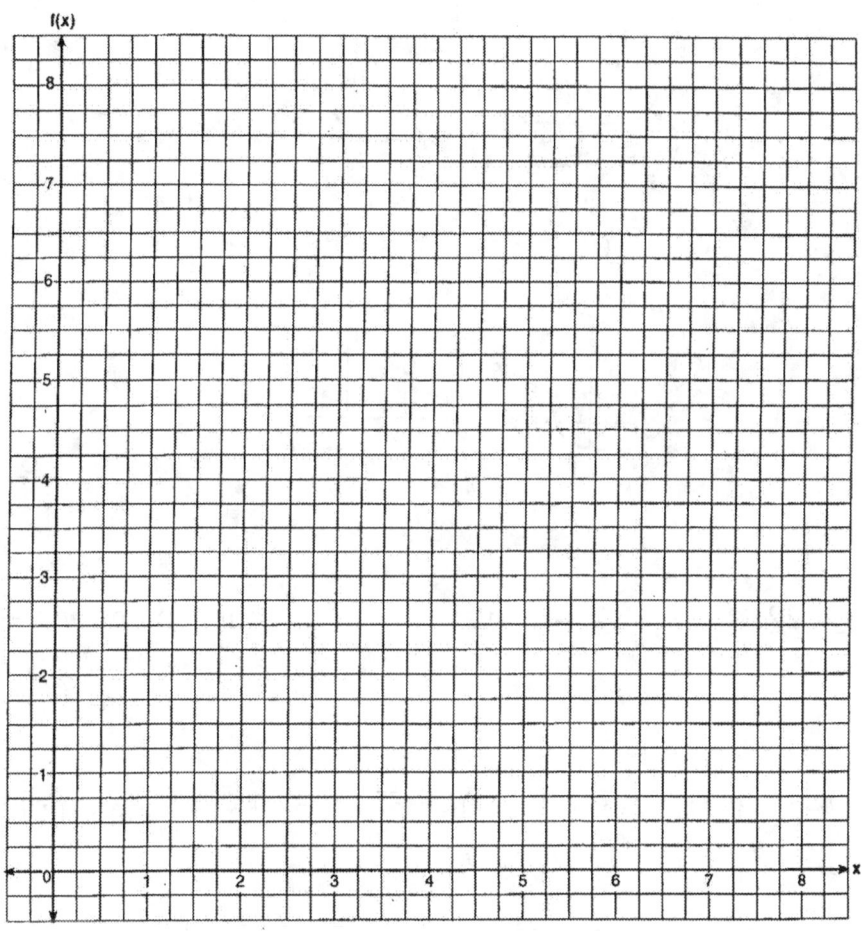

<u>June '03, #10:</u> Which diagram represents a relation in which each member of the domain corresponds to only one member of its range?

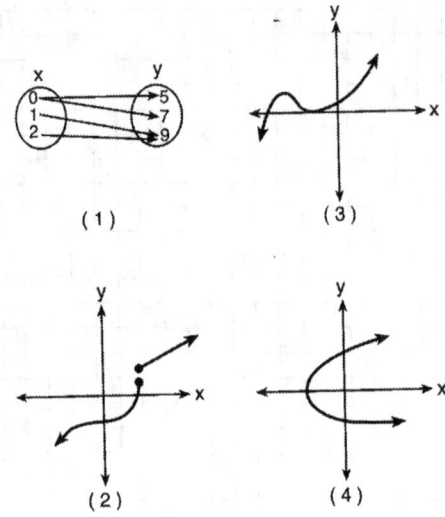

June '03, #22: If $f(x) = 2^x - 1$ and $g(x) = x^2 - 1$, determine the value of $(f \circ g)(3)$.

Jan '03, #31: If $f(x) = x^{\frac{2}{3}}$ and $g(x) = 8x^{-\frac{1}{2}}$, find $(f \circ g)(x)$ and find $(f \circ g)(27)$.

<u>Aug '03, #1:</u> Which graph does not represent a function of x?

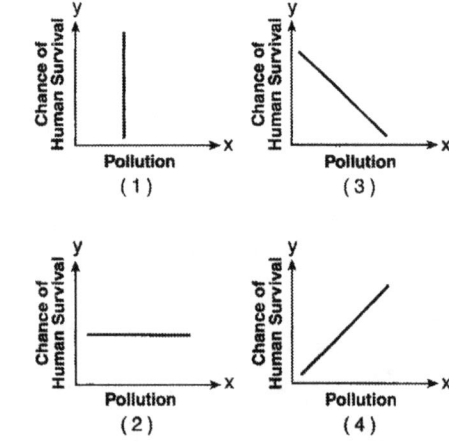

Aug '03, #13: If *f* and *g* are two functions defined by $f(x) = 3x + 5$ and $g(x) = x^2 + 1$, then $g(f(x))$ is

   (1) $x^2 + 3x + 6$ (2) $9x^2 + 30x + 26$ (3) $3x^2 + 8$ (4) $9x^2 + 26$

Aug '03, #19: A function is defined by the equation $y = \dfrac{1}{2}x - \dfrac{3}{2}$. Which equation defines the inverse of this function?

   (1) $y = 2x + 3$ (2) $y = 2x - 3$ (3) $y = 2x + \dfrac{3}{2}$ (4) $y = 2x - \dfrac{3}{2}$

Jan '04, #8: If $f(x) = \dfrac{2}{x+3}$ and $g(x) = \dfrac{1}{x}$, then $(g \circ f)(x)$ is equal to

   (1) $\dfrac{1+3x}{2x}$ (2) $\dfrac{2x}{1+3x}$ (3) $\dfrac{x+3}{2}$ (4) $\dfrac{x+3}{2x}$

Jan '04: #15: If $f(x) = x^3 - 2x^2$, then $f(i)$ is equivalent to

   (1) $-2 + i$ (2) $-2 - i$ (3) $2 + i$ (4) $2 - i$

Matching Models to Functions

<u>Sample #1:</u> The graph below represents the relationship of transported particle size to water velocity. The graph best models which type of function?

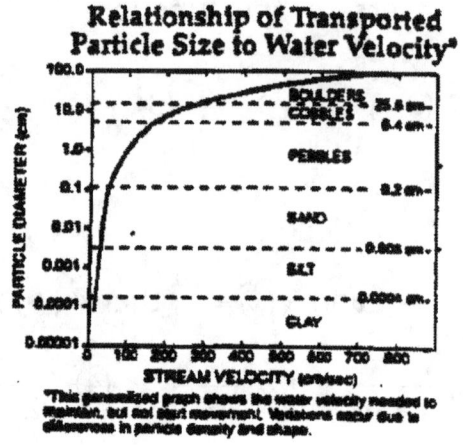

(1) linear   (2) quadratic   (3) logarithmic   (4) trigonometric

<u>Sample #2:</u> The graph below can be represented by which equation?

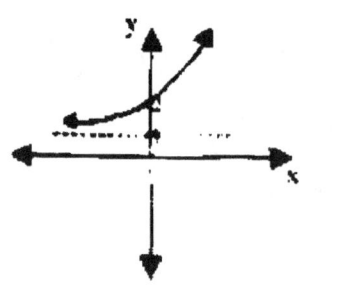

(1) $y = 2^x$ (2) $y = x^2 + 2$ (3) $y = 2^{x+1}$ (4) $y = 2^x + 1$

<u>Sample #13:</u> The price of a certain stock has decreased over 5 years, as shown in the graph below. Which of the following equations best represents this graph?

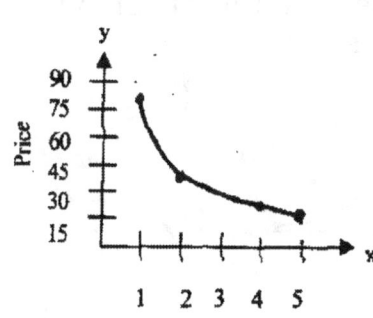

(1) $y = 60x^2$ (2) $y = \dfrac{80}{x}$ (3) $y = 63\log x$ (4) $y = -25x$

<u>Jan '02, #3:</u> The accompanying graph represents the value of a bond over time.

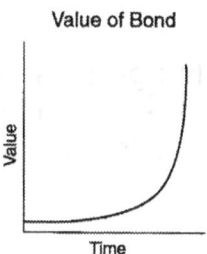

Which type of function does this graph best model?

(1) trigonometric   (2) logarithmic   (3) quadratic   (4) exponential

<u>June '03, #14:</u> Which equation is represented by the accompanying graph?

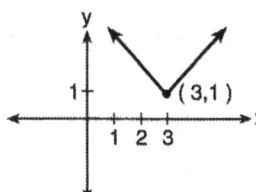

(1) $y = |x| - 3$ (3) $y = |x + 3| - 1$
(2) $y = (x - 3)^2 + 1$ (4) $y = |x - 3| + 1$

Aug '03, #4: The strength of a medication over time is represented by the equation $y = 200(1.5)^{-x}$, where x represents the number of hours since the medication was taken and y represents the number of micrograms per millimeter left in the blood. Which graph best represents this relationship?

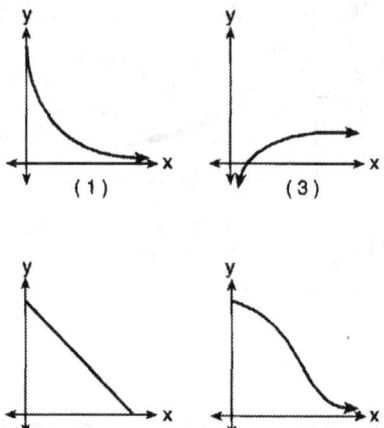

Aug '03, #12: The accompanying graph shows the relationship between a person's weight and the distance that the person must sit from the center of a seesaw to make it balanced.

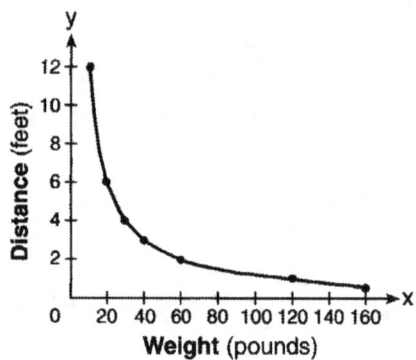

Which equation best represents this graph?

(1) $y = 12x^2$ (2) $y = -120x$ (3) $y = 2\log x$ (4) $y = \dfrac{120}{x}$

Law of Sines

The formula for the law of sines is found on the formula sheet for the Math B regents so it is not important to memorize it. ***It will be more important to recognize when it should be used.*** It is really three equations in one. This alone makes it difficult for the casual observer to choose the formula when it is needed.

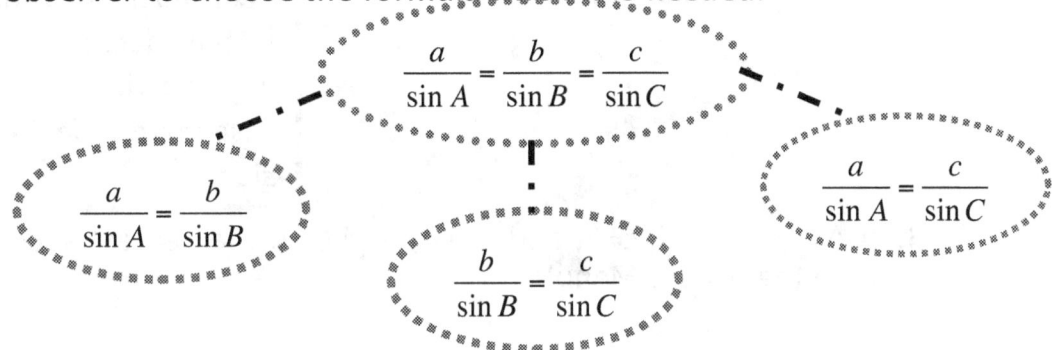

$$\frac{a}{\sin A} = \frac{b}{\sin B} = \frac{c}{\sin C}$$

$$\frac{a}{\sin A} = \frac{b}{\sin B}$$

$$\frac{b}{\sin B} = \frac{c}{\sin C}$$

$$\frac{a}{\sin A} = \frac{c}{\sin C}$$

These are just _____ .
To solve a proportion we need to know _____ .

To use the Law of Sines we need to know:

Remember: Proportions are solved by _____

Convincing ourselves that it should work:
Let's look at a right triangle.

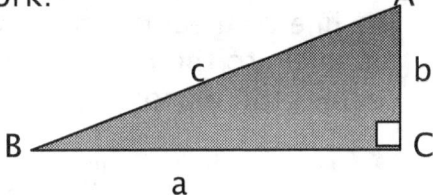

Sin A=	Sin B=	Sin C=	
c=	c=	c=	

Practice:

1. If C=82°, A=55°, and a=8, find c.

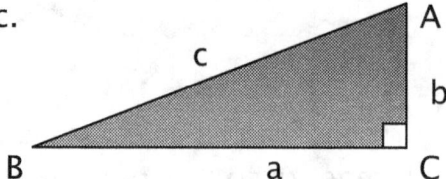

2. If A=45°, B=60°, and c=34, find a.

Remember: Capital letters denote angles and lower case letters denote sides.

3. If A=75°34'15", c=5, and a=10, find C to the nearest second. (Use the ANGLE Menu!)

4. The picture at the right depicts the great mathematician Archimedes setting fire to the Roman fleet as they attacked the city of Syracuse. If the distance between the two mirrors shown is 200 feet, the first is set at an angle of 74°45'14" from the line connecting the two and the second is set at an angle of 59°18'34" from the line connecting them, what is the distance, to the nearest foot, to the ship being ignited from each of the mirrors?

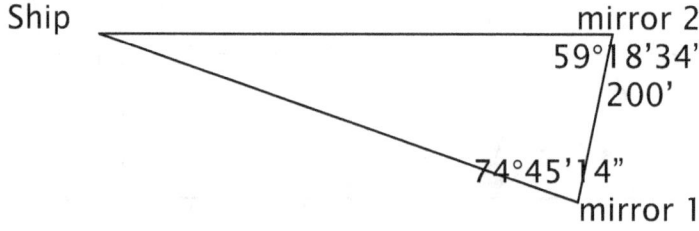

Ship   mirror 2
            59°18'34"
            200'

        74°45'14"
        mirror 1

5. Elaine and Sarah are walking in the woods. Elaine says she sees a bird's nest in a tree. She is looking at an angle of elevation of 23° and is 39 feet from the tree. Sarah says she sees it too. Sarah is looking at an angle of elevation of 31° and is 10 feet closer to the tree. Do Elaine and Sarah see the same nest? Explain.

One drawback of the Law of Sines is that we can sometimes have more than one correct answer or there could be no solution. Recall from our study of geometry that 2 sides and an angle only determine congruent triangles if the angle is between the known sides.

If we know two angles and one side we don't need to worry, there will be one solution.

If we are ***looking for an angle*** it's a little more complicated!

Most books give the cases according to the size of the known angle, but it may be easier to complete the Law of Sines to find the sine of the missing angle, then examine the result:

Case 1: The sine of the angle is greater than 1.	Case 2: The sum of the angle found with the given angle, or the sum of the supplement of the angle found with the given angle is ≥180, then there is one solution.	Case 3: If neither of the first two cases applies, there are two solutions.

The formula for the law of cosines is also found on the formula sheet. Again, it is very important to **know when to use it**.

$$a^2 = b^2 + c^2 - 2bc \cos A$$

To use the Law of Cosines we need to know:

The letters a, b, and c are interchangeable. Just be sure that the letter on the left is the lower case of the letter of the angle.

This formula is easy if one of the sides is unknown but we know the angle opposite that side.

If we are looking for the angle it will take a little more work.

First, an easy one:

A. In triangle ABC, b=20, c=23, A=119°. Find a.

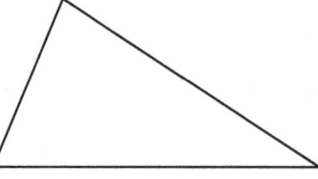

A little more work:

B. In triangle ABC, a=19, b=14, and c=12. Find C.

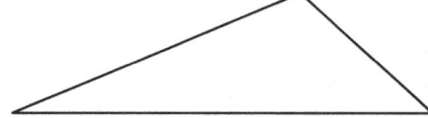

When it is the angle that is missing it is a good idea to check your answer with Equation Solver! There are too many places to make a computation error. Check your answer to part B. (Be sure to use parentheses correctly!)

Let's try a two-step!

C. In triangle ABC, b=6.2, c=7.8, and A=45°.   Find B.

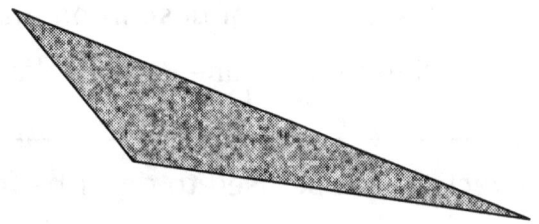

D. The lengths of the adjacent sides of a parallelogram are 54 cm and 78 cm.
   The larger angle measures 110°. What is the length of the longer diagonal? Round your answer to the nearest centimeter.

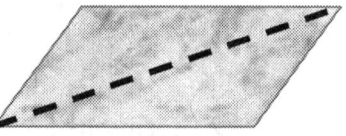

E. A boat runs in a straight line for 3 kilometers, then makes a 45° turn and goes for another 6 kilometers. How far is the boat from its starting point?

Vectors are ways of describing magnitude (length or size) and direction in the plane. (What plane are we talking about? _____)

Vectors can be used to describe force or velocity. Examples are:

The effect of two combined forces can be demonstrated using a parallelogram:

1. The two forces can be drawn as vectors that will form adjacent sides of the parallelogram.
2. Draw in the sides opposite the first two.
3. The result of their combined force (called the

   _____) is the diagonal drawn from

   their source to the opposite angle of the

   parallelogram.

4. The length of this vector is usually called the

   _____ .

5. When you are given an angle it is usually the one where the

   forces meet. Note that this is part of two triangles formed in

   the diagram. By itself it is not useful, but recall that in a

   parallelogram, adjacent angles are _____ .

6. Also recall that opposite sides of a parallelogram are

   _____ .

7. With these facts, when we are given the two forces and their included angle we can find the resulting force, or
8. Given the two forces and the resulting force, we can find the angle.
9. The most common "tool" to solve these problems is

   _____ .

Example:
  A. A river is flowing due east at 3 mph. A boat crosses the river headed due south. If the speedometer on the boat reads 4mph, what are the boat's actual speed and direction relative to the river bottom?

  B. If forces of 60 pounds and 20 pounds act on object such that the angle between them is 70°, what is the resultant force?

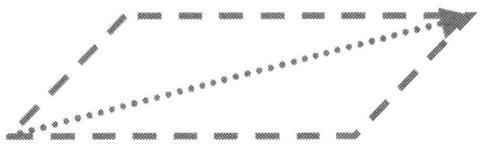

Area of a Triangle

We have always used the formula _____ to find the area of a triangle. To use this formula we needed to know _____ and _____ .

On the formula sheet provided for the Math B regents, we find the formula _____ labeled "Area of Triangle".

Are these the same thing?

$$(1/2)bh = (1/2)ab\sin C?$$

First we must recognize that, although both formulas contain "b", these "b's" are not the same. In our "old" formula, "b" is _____ . In the "new" formula, "b" is _____ . This will be a little confusing at first but we will soon be abandoning the "old" formula.

Let's look at triangle ABC. We will NOT assume that ABC is a right triangle. (Why?

_____
_____)

1. Begin by drawing in the altitude and labeling it "x". Remember, the altitude is _____ .
2. Suppose we know angle C. How could we find x?

_____

3. Rearrange the sine equation to solve for the altitude.
_____

4. Use the sine equivalent for the altitude and base "b" in our "old" area formula. _____

It really is the same formula!!

But now we can find the area of any triangle if we know

Practice:

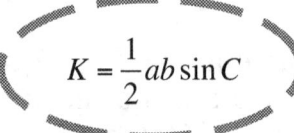
$$K = \frac{1}{2}ab\sin C$$

Reminder!!! What do we need to use this formula?

A. In triangle ABC, if a=4, b=3, and m∠C=55°, what is the area of the triangle?

B. In triangle ABC, if b=7, c=11, and m∠A=36°, what is the area of the triangle?

C. In triangle ABC, if a=22, b=34, m∠A=40°, and m∠B=48°, what is the area of the triangle?

D. A fence post is located 36 feet from one corner of a building and 40 feet from the adjacent corner. Fences are put up between the post and the building corners to form a triangular garden area. The 40-foot fence makes a 58° with the building. What is the area of the garden?

E. If Travis has a triangular garden with sides measuring 15 feet, 21 feet, and 30 feet, and Kyle has a triangular garden with sides measuring 16 feet, 24 feet, and 28 feet, whose garden has the larger area? How much larger is it to the nearest tenth of a square foot?

F. Melanie is making a quilt of triangular pieces of material. The finished sides of each triangle are 5 inches, 6.5 inches, and 8 inches. If Melanie needs the quilt to cover an area of 6 feet by 6 feet, what is the minimum number of pieces of material she will need to use?

G. An angle of a parallelogram has a measure of 145°. If the sides of the parallelogram measure 9 and 13 centimeters, what is the area of the parallelogram? (Hint: The parallelogram can be divided into 2 congruent triangles!)

Trigonometry:
Area, Law of Sines, Law of Cosines

<u>June '01, #21:</u> Gregory wants to build a garden in the shape of an isosceles triangle with one of the congruent sides equal to 12 yards. If the area of his garden will be 55 yards, find, to the nearest tenth of a degree, the three angles of the triangle.

<u>June '01, #27:</u> A wooden frame is to be constructed in the form of an isosceles trapezoid, with diagonals acting as braces to strengthen the frame. The sides of the frame each measure 5.30 feet, and the longer base measures 12.70 feet. If the angles between the sides and the longer base each measure 68.4°, find the length of one brace to the nearest tenth of a foot.

<u>Aug. '01, #8:</u> At Mogul's Ski Resort, the beginner's slope is inclined at an angle of 12.3°, while the advanced slope is inclined at an angle of 26.4°. If Rudy skis 1,000 meters down the advanced slope while Valerie skis the same distance on the beginner's slope, how much longer was the horizontal distance that Valerie covered?

(1) 81.3 m   (2) 231.6 m   (3) 895.7 m   (4) 977.0 m

<u>Aug. '01, #31:</u> In the accompanying diagram of triangle ABC, m∠A=65, m∠B=70, and the side opposite vertex B is 7. Find the length of the side opposite vertex A, and find the area of triangle ABC.

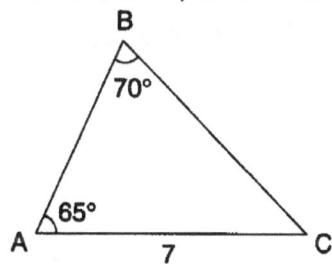

<u>Aug. '02, #18:</u> In the accompanying diagram of triangle ABC, m∠A=30, m∠C=50, and AC=13.

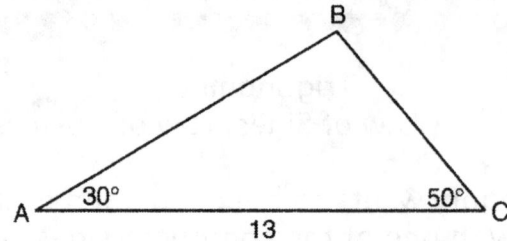

What is the length of side $\overline{AB}$ to the nearest tenth?

(1) 6.6   (2) 10.1   (3) 11.5   (4) 12.0

<u>Aug. '02, #26:</u> Two sides of a triangular-shaped pool measure 16 feet and 21 feet, and the included angle measures 58°. What is the area, to the nearest tenth of a square foot, of a nylon cover that would exactly fit the surface of the pool?

<u>Aug. '02, #33:</u> Carmen and Jamal are standing 5,280 feet apart on a straight, horizontal road. They observe a hot-air balloon between them directly above the road. The angle of elevation from Carmen is 60° and from Jamal is 75°. Draw a diagram to illustrate this situation and find the height of the balloon to the nearest foot.

<u>Jan. '02, #12:</u> In triangle ABC, m∠A=33, a=12, and b=15. What is m∠B to the nearest degree?

(1) 41   (2) 43   (3) 44   (4) 48

<u>Jan. '02, #25:</u> The accompanying diagram shows the floor plan for a kitchen. The owners plan to carpet all of the kitchen except the "work space," which is represented by scalene triangle ABC. Find the area of this work space to the nearest tenth of a square foot.

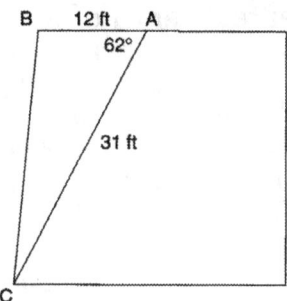

<u>June '02, #31:</u> A ship at sea heads directly toward a cliff on the shoreline. The accompanying diagram shows the top of the cliff, D, sighted from two locations, A and B, separated by distance S. If m∠DAC=30, m∠DBC=45, and S=30 feet, what is the height of the cliff, to the nearest foot?

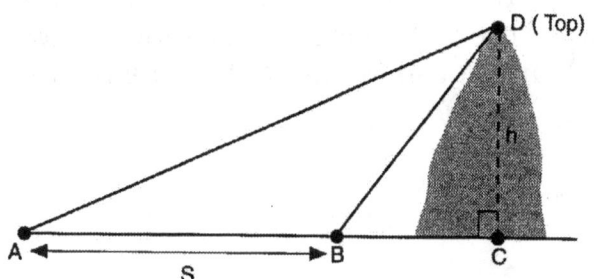

<u>June '02, #32:</u> Kieran is traveling from city A to city B. As the accompanying map indicates, Kieran could drive directly from A to B along County Route 21 at an average speed of 55 miles per hour or travel on the interstates, 45 miles along I-85 and 20 miles along I-64. The two interstates intersect at an angle of 150° at C and have a speed limit of 65 miles per hour. How much time will Kieran save by traveling along the interstates at an average speed of 65 miles per hour?

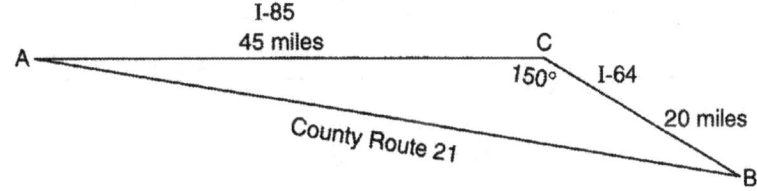

<u>Jan '02, #27:</u> Two straight roads, Elm Street and Pine Street, intersect creating a 40° angle, as shown in the accompanying diagram. John's house (J) is on Elm Street and is 3.2 miles from the point of intersection. Mary's house (M) is on Pine Street and 5.6 miles from the intersection. Find, to the nearest tenth of a mile, the direct distance between the two houses.

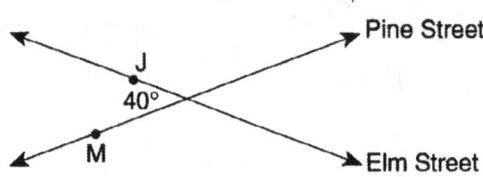

<u>Sample #29:</u> The Vietnam Veteran's Memorial in Washington, DC consists of two walls of black polished granite, each 246.75 feet long, which meet at an angle of 125.2°. If extended, the west wall would reach to the Lincoln Memorial, 900 feet away from the end of the wall and the east wall would reach to the Washington Monument 3,500 feet away from the end of the wall. Find the distance between the Lincoln Memorial and the Washington Monument to the nearest foot.

<u>June '03, #32:</u> While sailing a boat offshore, Donna sees a lighthouse and calculates that the angle of elevation to the top of the lighthouse is 3°, as shown in the accompanying diagram. When she sails her boat 700 feet closer to the lighthouse, she finds that the angle of elevation is now 5°. How tall, to the nearest foot, is the lighthouse?

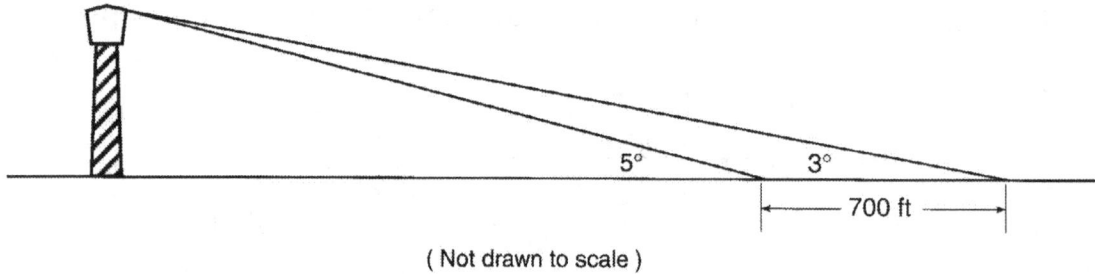

(Not drawn to scale)

June '03, #33: A farmer has determined that a crop of strawberries yields a yearly profit of $1.50 per square yard. If strawberries are planted on a triangular piece of land whose sides are 50 yards, 75 yards, and 100 yards, how much profit, to the nearest hundred dollars, would the farmer expect to make from this piece of land during the next harvest?

Jan '03, #9: In ΔABC, if AC=12, BC=11, and m∠A=30, angle C could be

    (1) an obtuse angle, only
    (2) an acute angle, only
    (3) a right angle, only
    (4) either an obtuse angle or an acute angle

Jan '03, #30: A picnic table in the shape of a regular octagon is shown in the accompanying diagram. If the length of $\overline{AE}$ is 6 feet, find the length of one side of the table to the nearest tenth of a foot, and find the area of the table's surface to the nearest tenth of a square foot.

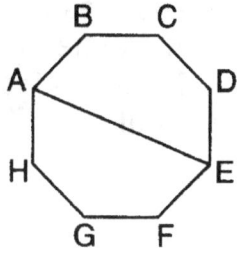

Jan '03, #34: A ship captain at sea uses a sextant to sight an angle of elevation of 37° to the top of a lighthouse. After the ship travels 250 feet directly toward the lighthouse, another sighting is made, and the new angle of elevation is 50°. The ship's charts show that there are dangerous rocks 100 feet from the base of the lighthouse. Find, to the nearest foot, how close to the rocks the ship is at the time of the second sighting.

Aug '03, #11: An architect commissions a contractor to produce a triangular window. The architect describes the window as $\triangle ABC$, where $m\angle A = 50$, BC=10 inches, and AB=12 inches. How many distinct triangles can the contractor construct using these dimensions?

(1) 1   (2) 2   (3) more than 2   (4) 0

Aug '03, #24: The triangular top of a table has two sides of 14 inches and 16 inches, and the angle between the sides is 30°. Find the area of the tabletop, in square inches.

Aug '03, #29: A ship at sea is 70 miles from one radio transmitter and 130 miles from another. The angle between the signals sent to the ship by the transmitters is 117.4°. Find the distance between the two transmitters, to the nearest mile.

Jan '04, #7: In $\triangle ABC$, a=19, c=10, and $m\angle A = 111$. Which statement can be used to find the value of $\angle C$

(1) $\sin C = \dfrac{10}{19}$

(2) $\sin C = \dfrac{19\sin 69°}{10}$

(3) $\sin C = \dfrac{10\sin 21°}{19}$

(4) $\sin C = \dfrac{10\sin 69°}{19}$

Jan '04, #17: A garden in the shape of an equilateral triangle has sides whose lengths are 10 meters. What is the area of the garden?

(1) $25m^2$ (2) $25\sqrt{3}m^2$ (3) $50m^2$ (4) $50\sqrt{3}m^2$

Jan '04, #26: A landscape designer is designing a triangular garden with two sides that are 4 feet and 6 feet respectively. The angle opposite the 4-foot side is 30°. How many distinct triangular gardens can the designer make using these measurements?

Aug '02, #7: To balance a seesaw, the distance, in feet, a person is from the fulcrum is inversely proportional to the person's weight, in pounds. Bill, who weighs 150 pounds, is sitting 4 feet away from the fulcrum. If Dan weighs 120 pounds, how far from the fulcrum should he sit to balance the seesaw?

(1) 4.5 ft   (2) 3.5 ft   (3) 3 ft   (4) 5 ft

Aug '02, #28: Two tow trucks try to pull a car out of a ditch. One tow truck applies a force of 1,500 pounds, while the other truck applies a force of 2,000 pounds. The resultant force is 3,000 pounds. Find the angle between the two applied forces, rounded to the nearest degree.

Jan '04, #30: One force of 20 pounds and one force of 15 pounds act on a body at the same point so that the resultant force is 19 pounds. Find, to the nearest degree, the angle between the two original forces.

Trigonometry: Double Angles and Half Angles
Sums and Differences of Angles

<u>June '02, #22:</u> Is $\frac{1}{2}\sin 2x$ the same expression as $\sin x$? Justify your answer.

<u>Sample #5:</u> If x is an acute angle, and $\cos x = \frac{4}{5}$, then $\cos 2x$ is equal to

(1) $\frac{6}{25}$ (2) $\frac{-1}{25}$ (3) $\frac{2}{25}$ (4) $\frac{7}{25}$

<u>June '01, #31:</u> In the interval $0° \le A < 360°$, solve for all values of A in the equation $\cos 2A = -3\sin A - 1$.

<u>June '01, #18:</u> If θ is an obtuse angle and $sin\ \theta = b$, then it can be concluded that
    (1) $tan\ \theta > b$ (3) $cos\ 2\theta > b$
    (2) $cos\ \theta > b$ (4) $sin\ 2\theta < b$

<u>Jan '03, #19:</u> If $\sin\theta = \frac{\sqrt{5}}{3}$, then $\cos 2\theta$ equals

(1) $\frac{1}{3}$ (2) $-\frac{1}{3}$ (3) $\frac{1}{9}$ (4) $-\frac{1}{9}$

<u>Aug '03, #15:</u> The expression $\frac{2\cos\theta}{\sin 2\theta}$ is equivalent to

(1) $\csc\theta$ (2) $\sec\theta$ (3) $\cot\theta$ (4) $\sin\theta$

<u>Aug '03, #16:</u> If $\sin x = \dfrac{12}{13}$, $\cos y = \dfrac{3}{5}$, and x and y are acute angles, the value of $\cos(x - y)$ is

(1) $\dfrac{21}{65}$ (2) $\dfrac{63}{65}$ (3) $-\dfrac{14}{65}$ (4) $-\dfrac{33}{65}$

<u>Jan '04, #1:</u> The expression $\cos 40° \cos 10° + \sin 40° \sin 10°$ is equivalent to

(1) $\cos 30°$ (2) $\cos 50°$ (3) $\sin 30°$ (4) $\sin 50°$

<u>Jan '04, #18:</u> If x is an acute angle and $\sin x = \dfrac{12}{13}$, then $\cos 2x$ equals

(1) $\dfrac{25}{169}$ (2) $\dfrac{119}{169}$ (3) $-\dfrac{25}{169}$ (4) $-\dfrac{119}{169}$

Trigonometry: Solving Trigonometric Equations
Identities

<u>June '02, #33:</u> On a monitor, the graphs of two impulses are recorded on the same screen, where $0° \le x < 360°$. The impulses are given by the following equations:

$$y = 2\sin^2 x$$
$$y = 1 - \sin x$$

Find all values of x, in degrees, for which the two impulses meet in the interval $0° \le x < 360°$. [Only an algebraic solution will be accepted.]

<u>June '03, #19:</u> What value of x in the interval $0° \le x \le 180°$ satisfies the equation $\sqrt{3}\tan x + 1 = 0$?

(1) -30°   (2) 30°   (3) 60°   (4) 150°

<u>Jan '03, #17:</u> If $(\sec x - 2)(2\sec x - 1) = 0$, then x terminates in

(1) Quadrant I, only
(2) Quadrants I and II, only
(3) Quadrants I and IV, only
(4) Quadrants I, II, III, and IV

<u>Jan '03, #20**:</u> If $\sin 6A = \cos 9A$, then $m\angle A$ is equal to

(1) 6   (2) 36   (3) 54   (4) $1\dfrac{1}{2}$

<u>Jan '04, #2:</u> The expression $\dfrac{\sec\theta}{\csc\theta}$ is equivalent to

(1) $\sin\theta$ (2) $\cos\theta$ (3) $\dfrac{\sin\theta}{\cos\theta}$ (4) $\dfrac{\cos\theta}{\sin\theta}$

<u>Jan '04, #4:</u> What is a positive value of x for which $9^{-\cos x} = \dfrac{1}{3}$?

(1) 30° (2) 45° (3) 60° (4) 90°

$\sum$ Some Sum!

In math we sometimes want to do the same computation on a series of numbers and find their total. To show this we use

_____ _____

and use the symbol _____ _____ or $\sum$.

Example:

$$\sum_{n=2}^{5} 3n - 1$$

The letter below sigma shows _____

The number below sigma tells _____

The number above sigma tells _____

The expression after sigma _____

On the calculator we can enter this as the sum of a sequence:
1. Press 2ⁿᵈ STAT
2. Go over to MATH
3. Choose 5: sum(
4. Press 2ⁿᵈ STAT again
5. Go over to OPS
6. Choose 5: seq(
7. Enter the "rule" or expression following the summation symbol, the variable from the expression, the lower number, and the higher number all separated by commas.
8. Press ENTER.

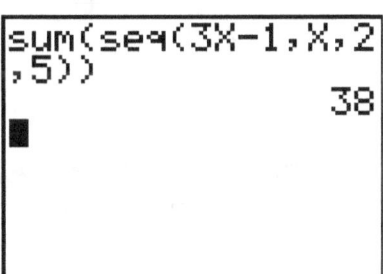

Practice:

a. $\displaystyle\sum_{a=1}^{4} 2a$ _____

b. $\displaystyle\sum_{i=3}^{6} 4i - 3$ _____

a+bi mode won't work with this method; this one will need to be done the "long way".

c. $\displaystyle\sum_{x=3}^{7} 2x - i$ _____

Note the difference between b and c! "*i*" is a variable in b. It is the imaginary unit in c.

d. $\displaystyle\sum_{x=3}^{8} 2x$ _____

e. $\displaystyle\sum_{k=1}^{4} 2k^2$ _____

f. $\displaystyle\sum_{k=1}^{5} (k-1)^3$ _____

Place the ½ before "sum" on the home screen!

g. $\displaystyle\frac{1}{2}\sum_{x=1}^{4} (x-1)^2$ _____

h. $\displaystyle\frac{2}{3}\sum_{n=-1}^{6} (3x-9)$ _____

i. $\displaystyle 3\sum_{x=1}^{5} x^{x-2}$ _____

j. $\displaystyle\frac{2}{5}\sum_{x=-2}^{2} (2x-7)^{3x}$ _____

Summation and Series

June '02, #1: What is the value of $\sum_{m=2}^{5}(m^2-1)$?

    (1) 58   (2) 54   (3) 53   (4) 50

June '01, #17: What is the value of $\sum_{m=1}^{3}(2m+1)^{m-1}$?

    (1) 15   (2) 55   (3) 57   (4) 245

June '03, #26: Evaluate: $2\sum_{n=1}^{5}(2n-1)$

Aug '02, #13: If $_nC_r$ represents the number of combinations of n items taken r at a time, what is the value of $\sum_{r=1}^{3}{}_4C_r$?

    (1) 24   (2) 14   (3) 6   (4) 4

Jan '03, #4: What is the value of $\sum_{b=0}^{3}(2-(b)i)$?

    (1) $2-5i$ (2) $2-6i$ (3) $8-5i$ (4) $8-6i$

Introduction to Programming

Programming the TI-83+/TI-84+ could be a complete course itself. We will do just enough here to demonstrate how awesome and adaptable this tool can be.

The TI-83+/TI-84+ is not really a calculator; it is really a hand-held computer that we can program to do almost anything that isn't already built in.

First, we will have two types of commands: control commands and input/output instructions. Below are listed only those that will be used in the two programs included in this lesson.

Control Commands: (PRGM, CTR)
1. <u>If</u>: Creates a conditional test. Passing the test takes it to the next command. Failing will cause it to ignore the command immediately following and go on with the next command.
2. <u>Lbl</u>: Defines a label. Used to give the program a place to "jump" to from an "if" statement.
3. <u>Goto</u>: Goes to a label. Usually follows an "if" test.
4. <u>Stop</u>: Stops execution. Pressing ENTER will return to the beginning of the program when this command is used. It is optional.

Input/Output Commands: (PRGM, I/O)
1. <u>Input</u>: Creates a place for a value to be entered.
2. <u>Disp</u>: Sends text or a value to the screen to be viewed during execution.
3. <u>ClrHome</u>: Clears the display so that the steps in the execution of the program are the only entries visible on the screen.
4. <u>Prompt</u>: Gives a place for data to be entered and assigned to a variable in the program.

To get started, press PRGM and go over to NEW. Press ENTER to choose 1:Create New.

You will be prompted to name your new program. A program name can have 1 to 8 characters that are letters, numbers, or θ. The only restriction is that it cannot begin with a number.

```
EXEC EDIT NEW
1:Create New
```

Our first program will find the roots of any quadratic equation.

1. Press PRGM, go over to NEW and press ENTER to choose 1:Create New.
2. Call the new program QUADROOT.
3. Press PRGM, go over to I/O and choose 8:ClrHome. ENTER
4. Press PRGM, go over to I/O and choose 3:Disp, "AX²+BX+C=0". ENTER
5. Press PRGM, go over to I/O and choose 2:Prompt, A. ENTER

6. Press PRGM, go over to I/O and choose 2:Prompt, B. ENTER
7. Press PRGM, go over to I/O and choose 2:Prompt, C. ENTER
8. Enter (B²-4AC) stored to N. (Calculations do not need a command.)
9. Press PRGM, choose 1:If N, 2ⁿᵈ MATH, choose 5:<, 0. ENTER
10.   Press PRGM, choose 0:Goto, 1. ENTER
11.   Press PRGM, go over to I/O, choose 3:Disp, (-B+√N)/(2A), ENTER
12.   Press PRGM, go over to I/O, choose 3:Disp, (-B-√N)/(2A), ENTER
13.   Press PRGM, choose F:Stop, ENTER
14.   Press PRGM, choose 9:Lbl, 1, ENTER
15.   Press PRGM, go over to I/O, choose 3: Disp, (-B+√(abs(N))i)/(2A), ENTER
16.   Press PRGM, go over to I/O, choose 3: Disp, (-B-√(abs(N))i)/(2A), ENTER

To execute the program QUIT, press PRGM and go over to EXEC.

Note: Do not choose EXEC without quitting the new program or edit menu. The EXEC command located here allows you to execute a program from within another program and is not the one we will usually want.

Follow the directions given on the screen.

Use your program to find the roots of the following quadratic equations.

1. $y = 2x^2 - 3x + 1$ _____ _____

2. $y = -3x^2 + 4x + 7$ _____ _____

3. $y = 5x^2 + 11x - 4$ _____ _____

4. $y = -x^2 + 7x - 14$ _____ _____

5. $y = 11x^2 + 3x + 10$ _____ _____

A program could be very useful on the job. If you have a calculation that you use frequently it can be entered as a program to make your job easier.

For example: Suppose Andrea is working at the window of a soft ice cream shop. If sundaes cost $1.75 and cones cost $1.25 and there is 8% sales tax on each sale, the following program could be used to calculate the cost and the amount of change due to each customer.

Program Name: ICECREAM
: ClrHome
: Disp "HOW MANY SUNDAES?"
: Input S
: Disp "HOW MANY CONES?"
: Input C
: Disp "AMOUNT DUE"
: (S*1.75+C*1.25)*1.08□X
: Disp X
: Disp "AMOUNT PAID"
: Input A
: Disp "CHANGE DUE"
: Disp (A-X)
: Stop

Note: If you notice that your menu choices are CTL, I/O, and EXEC but you want to edit the program, you will need to quit then return to the program menu.

Use the program ICECREAM to find the following:

1. The Noftsier's order two sundaes and two cones. What do they owe? _____ . They pay with a ten-dollar bill, what is their change? _____

2. The Yancey's order 4 cones and 3 sundaes. What do they owe? _____ . They pay with a twenty-dollar bill, what is their change? _____ .

3. The Houppert's order 5 cones. What do they owe? _____. They pay with a twenty-dollar bill, what is their change?

A program can be changed by choosing EDIT in the PRGM menu. Use INSERT and DELETE to add or take out parts of the program.

A program will be lost if a Reset All is chosen. A program can be archived to protect it from accidental deletion.

Programs can be transferred from one calculator to another, or from your calculator to a computer.

More programs can be found in the TI-83+ manual, on the Internet, and in some math textbooks.

Review 1

1. Write the quadratic formula: _____
2. Write the discriminant formula: _____
3. Write the formula for the axis of symmetry: _____
4. If $b^2 - 4ac < 0$ then the roots of the equation $ax^2 + bx + c = 0$ must be
   - a. Real, irrational, and unequal
   - b. Real, rational, and unequal
   - c. Real, rational, and equal
   - d. Imaginary _____
5. The roots of the equation $x^2 + 7x - 8 = 0$ are
   - a. Real, irrational, and unequal
   - b. Real, rational, and unequal
   - c. Real, rational, and equal
   - d. Imaginary _____
6. The roots of the equation $ax^2 + 4x = -2$ are real and equal if a has a value of
   - a. 1   b. 2   c. 3   d. 4   _____

7. Evaluate $y = (16x)^0 + x^{\frac{2}{3}}$ when $x=64$. _____

8. Find the value of $x^{-\frac{3}{2}}$ if $x=16$. _____

9. Evaluate: $4^0 - 8^{\frac{2}{3}} + 9^{\frac{1}{2}}$ _____

10. Solve: $\sqrt{x^2 - 5x + 5} = 1$

_____

11. The expression $\dfrac{2 + \sqrt{3}}{2 - \sqrt{3}}$ is equivalent to

   - a. $11\sqrt{3}$ b. $7 - 4\sqrt{3}$ c. $7 + 4\sqrt{3}$ d. $\dfrac{7 + 4\sqrt{3}}{7}$

_____

12. If $4\sqrt{5} = \sqrt{n}$, the value of n is
   - a. 10   b. 20   c. 80   d. 100   _____

13. Solve for x: $\sqrt{x + 10} + 2 = 5$

_____

14. Find the positive value of x if $|4 - 2x| = 6$ _____

15. Find the turning point and axis of symmetry for the equation $y = -2x^2 + 6x - 7$. _____ _____

16. Find the turning point and axis of symmetry for the equation $y = 3x^2 + 7x - 2$. _____ _____

17. A stone is dropped from a high bridge and follows a path that can be modeled by the equation $h(x) = -4.9x^2 + 2x + 50$, where h is the height above the water in meters and x is time in seconds from when the stone is dropped.
   a. Graph the equation on the grid below.
   b. If a camera takes a picture 1.2 seconds after the stone is dropped, at what height above the water will the stone be in the picture? _____
   c. To the nearest tenth of a second, how many seconds after the stone is dropped will it hit the water? _____

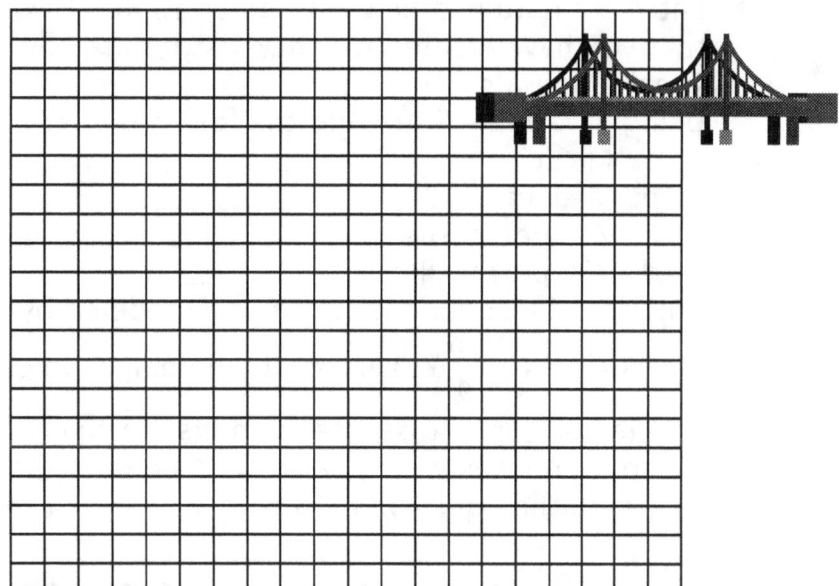

18. Graph $y = -|3x + 1| + 5$

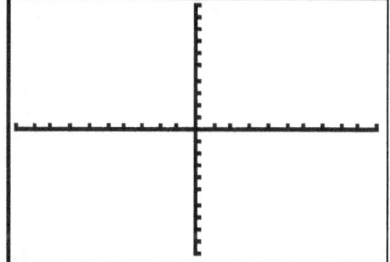

   a. Rewrite the equation so that the slope is less steep.

   _____

   b. Rewrite the equation so that the turning point is on the x-axis.

   _____

   c. Rewrite the equation so that the graph is turned "upside-down" from its original position. _____

19.   At a frog jumping contest, 9 frogs were entered and had the jumps recorded at the right:

Frog	Length of Jump
Kermy	19.4cm
Legs	20.3cm
Hopper	25.9cm
L'il Leaper	11.0cm
Freddie	22.6cm
Fidget	20.8cm
Slimy	16.1cm
Croaker	14.7cm
Chomper	20.0cm

Enter the data in a new list named JUMP. Have the teacher verify that you have created this new list before continuing.

Teacher Initials: _____

Using 1-Variable Stats find the following:

    a. Mean: _____

    b. Median: _____

    c. 25th percentile: _____

    d. 75th percentile: _____

20.   If Hopper's jump can be modeled by the equation

$y = -.3x^2 + \dfrac{2000}{259}x$, where y is the height in centimeters and x is

the horizontal distance from his starting point in centimeters, how high was he at the highest point in his leap? _____

Graph the equation below.

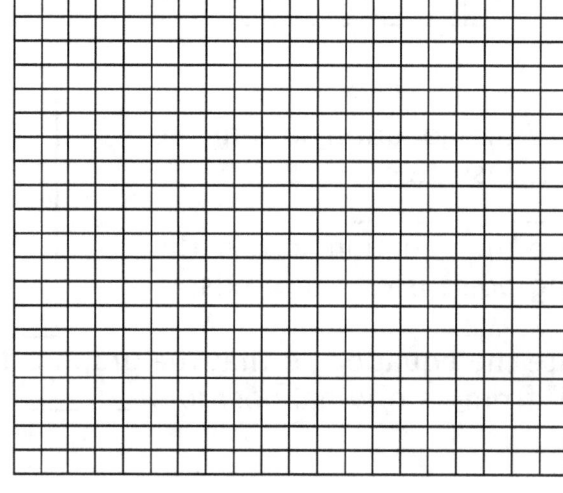

21.   Graph and find the solution set:
$x^2 - 3x - 28 \geq 0$

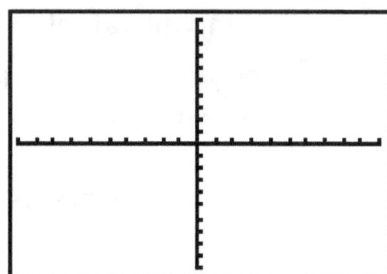

----

22.   Graph and find the solution set:
$x^2 - 4x - 5 < 0$

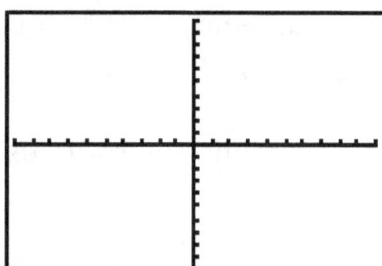

----

23.   Graph and find the solution set:
$|3x - 2| < 4$

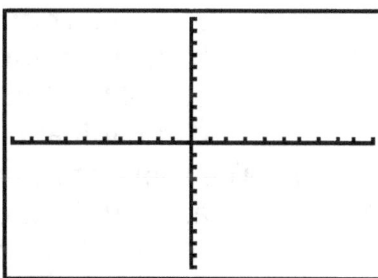

----

24.   Graph and find the solution set:
$|3x - 4| > 5$

Review 2

1. Solve for x: $6-\sqrt{x-2}=2$ _____

2. Find all values of x: $|3x-7|=20$ _____

3. Find the turning point and axis of symmetry for the equation $y=-x^2+11x-7$.

          _____   _____

4. Find the roots using the quadratic formula: $0=-x^2+11x-7$

          _____

5. Find the turning point and axis of symmetry for the equation $y=2x^2-3x-2$.

          _____   _____

6. Find the roots using the quadratic formula: $0=2x^2-3x-2$

          _____

7. The roots of the equation $-2x^2+5x-8=0$ are
   a. Real, irrational, and unequal
   b. Real, rational, and unequal
   c. Real, rational, and equal
   d. Imaginary   _____

8. The roots of the equation $x^2+4x+4=0$ are
   a. Real, irrational, and unequal
   b. Real, rational, and unequal
   c. Real, rational, and equal
   d. Imaginary   _____

9. The roots of the equation $x^2+8x-8=0$ are
   a. Real, irrational, and unequal
   b. Real, rational, and unequal
   c. Real, rational, and equal
   d. Imaginary   _____

10. The roots of the equation $4x^2+6x-4=0$ are
    a. Real, irrational, and unequal
    b. Real, rational, and unequal
    c. Real, rational, and equal
    d. Imaginary   _____

11. Graph and find the solution set:
$2x^2 + 3x - 5 \le 4$

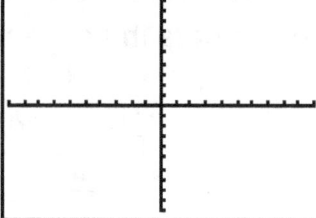

12. Graph and find the solution set:
$-|3x - 5| + 7 \ge -2$

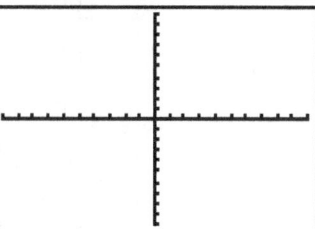

13. Graph $y = |3x - 7| + 4$

14. Change the equation in #13 so that the turning point lies on the x-axis.

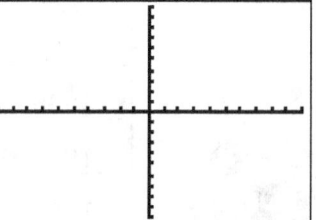

_____

15. A tennis ball machine has its ball velocity set at 45 mph. The path of the ball can be described by the function $h = -16s^2 + 66s + 2$, where h is the height of the ball in feet and s is time in seconds.
   a. Graph the function on the grid below.
   b. What is the maximum height of the tennis ball?
   c. How many seconds after being fired will the ball land?

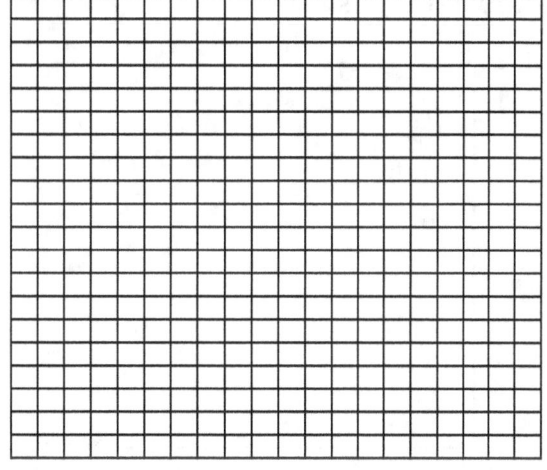

16. If the ball machine is 2 feet high, a ball reaches a maximum height of 20 feet after 3 seconds and lands 5 seconds after being fired from the machine, what is the initial velocity of the ball in feet per second? _____

Use the data in the table below to answer questions #17-23.

US Snowmobile Sales:

Year	Units	Dollars
2003	114927	779246951
2002	134082	817331445
2001	140629	894359351
2000	136601	821000000
1999	147867	882766000
1998	162826	975147000
1997	170325	1005790000
1996	168509	905194000
1995	148207	791277000
1994	114057	556879000
1993	87809	403921000
1992	81946	356000000

17.   Create a scatter plot with years since 1992 as the independent variable and number of snowmobiles sold as the dependent variable. Sketch below being sure to indicate window dimensions.

18. Fill in the table below and draw a box-and-whisker plot describing the number of snowmobiles sold per year from 1992 to 2003.

Mean	
Minimum	
25th Percentile	
Median	
75th Percentile	
Maximum	

19. Find the line of best fit for units sold vs. dollars.

_____

Is there a strong linear relationship? _____ Explain using the correlation coefficient.

_____

_____

20. Using 2-Variable statistics find:

a. Mean # of units sold: _____

b. Total # of units sold: _____

c. Mean # of dollars in sales: _____

d. Total # of dollars in sales: _____

21. Find the variance in the number of units sold: _____

22. Find the variance in the number of dollars in sales:

_____

23. What is unusual about the data?

_____

_____

24. In January 1998 many maple trees were damaged by an ice storm. New York Agricultural Statistics conducted a survey to assess the extent of the damage. The table below indicates the responses of the 63% of maple producers that responded to the survey.

Maple Producers Reporting Damage			
County	Sugarbushes (Percent by county)	Percent reporting Heavy Damage	Percent of Taps Not Placed
Clinton	99	90	89
Essex	57	42	57
Franklin	94	60	80
Jefferson	79	70	86
Lewis	30	29	26
St. Lawrence	94	68	77
Wyoming	44	25	13

a. Find a power model comparing percent of sugarbushes with percent reporting heavy damage. _____
Is it a good model? _____
Explain._____

_____

b. Find a power model comparing percent of sugarbushes reporting heavy damage with percent of taps not placed.

_____
Is it a good model? _____
Explain.

_____

_____

c. Is a linear model better for either or both of the comparisons

above? _____

Explain.

_____

_____

Review 3

1. Find the discriminant of the equation $0 = 4x^2 - 3x + 7$.

   ----

2. The roots of $0 = 2x^2 - 3x - 9$ are
   a. Real, rational, and equal
   b. Real, rational, and unequal
   c. Real, irrational, and unequal
   d. Imaginary

   ----

3. The roots of $0 = 5x^2 - 9x + 4$ are
   a. Real, rational, and equal
   b. Real, rational, and unequal
   c. Real, irrational, and unequal
   d. Imaginary

   ----

4. Use the quadratic formula to find the roots of the equation $2x^2 - 7x - 4 = 0$.

   ----

5. Solve for x: $|3x - 7| + 4 = 12$

   ----

6. Solve for x: $\sqrt{3x + 15} = 6x$

   ----

7. Graph and sketch: $y = -|4x + 1| - 2$

   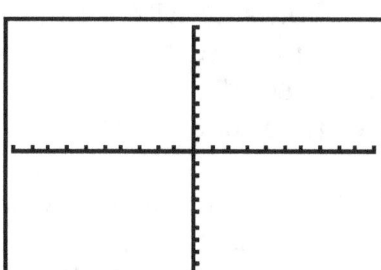

8. Find the turning point and axis of symmetry for the equation $y = 3x^2 - 3x + 1$.

   -------- ------------

9. Find the turning point and axis of symmetry of the equation $y = -x^2 + 5x - 7$.

10. Graph and find the solution set: $5 \le -x^2 + 2x + 7$

11. Graph and find the solution set: $|2x - 6| \ge 3$

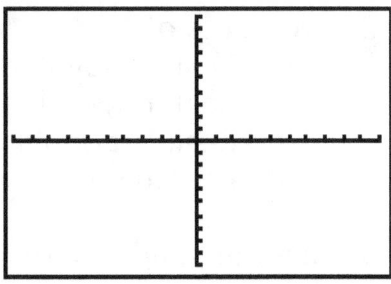

12. In 1896 the Olympic Marathon winner was Spiridon Louis of Greece with a time of 2 hours, 58 minutes, and 50 seconds. In 2000, the winner was Gezahgne Abera of Ethiopia with a time of 2 hours, 10 minutes, and 11 seconds. How much faster was Gezaghne's time than Spiridon's?

13. At the 2000 Summer Olympics the women's gold medallist in the 100 meter backstroke was Diana Mocanu from Romania with a time of 1 minute and .2 seconds. The men's winner was Lenny Krayzlburg of the USA with a time of 53.72 seconds. How much longer did it take Diana to complete the race than Lenny?

14. The St. Louis Arch is 630 feet high and 630 feet wide measuring from the outside edges of the bases. If the edge of one base is at the origin, write a quadratic formula that would graph the arch to scale.

15. A basketball player releases the ball at mid-court from a height of 7'. The ball reaches a maximum height of 20' and goes in the basket that is 10' off the ground. If the court is 54' long, write a quadratic equation that models the ball's path.

State	Governor's Salary	Attorney General's Salary
Connecticut	$150,000	$81,562
Delaware	114,000	114,400
Maine	70,000	78,000
Maryland	120,000	100,000
Massachusetts	135,000	122,500
New Hampshire	100,690	85,753
New Jersey	157,000	137,165
New York	179,000	151,500
Ohio	126,000	93,434
Pennsylvania	144,416	118,262
Rhode Island	95,000	85,000
Vermont	88,026	90,272

16. Using the table above, find the following:
a. Variance in governors' salaries: _____
b. Mean Absolute deviation in governors' salaries:

_____

c. Standard deviation in attorney generals' salaries:

_____

17. If the salaries for all of the state governors' were normally distributed with a mean of 111333.62 and a standard deviation of 26549.93, what percent of the governors would have a salary greater than $150,000?

----

18. How many governors would have a salary less than $100,000?

Population in 1000's

Year	New York	California	United States
1790	340		3929
1800	589		5308
1810	959		7240
1820	1373		9638
1830	1919		12866
1840	2429		17069
1850	3097	93	23192
1860	3881	380	31443
1870	4383	560	39818
1880	5083	865	50156
1890	6003	1213	62948
1900	7269	1485	75995
1910	9114	2378	91972
1920	10385	3427	105711
1930	12588	5677	122775
1940	13479	6907	131669
1950	14830	10586	150697
1960	16782	15717	179323
1970	18237	19953	203302
1980	17558	23668	226546
1990	17991	29811	248791
2000	18976	33872	281422

19. Using the data above, create scatter plots with 1790 as the zero year.

New York	California

```
┌─────────────────────────────────────────┐
│              United States              │
│                                         │
│                                         │
│                                         │
│                                         │
│                                         │
│                                         │
│                                         │
└─────────────────────────────────────────┘
```

20. Find the best regression model for each set of data in the table on the previous page. (Linear, exponential, logarithmic, or power.)
    **Omit years there is no data for California when finding the model for California. Use .01 in place of the 0 year if a domain error occurs.

    New York: _____ r= _____

    California: _____ r= _____

    United States: _____ r= _____

21. Using the best model, approximate the population for the year 2020.

    New York: _____

    California: _____

    United States: _____

    Why does the data for California start at 1850 rather than 1790?

    _____

    _____

    Based on the intervals in the data, what is the likely source of

    the population data?

    _____

    _____

22. The table below shows maple sap collected on a particular day.
   a. Find the total amount of sap collected. _____
   b. Find the variance. _____
   c. Find the mean absolute deviation. _____
   d. Find the standard deviation. _____

Amount of sap in quarts	# of buckets
0.5	3
1.0	4
1.5	6
2.0	8
2.5	5
3.0	10
3.5	3
4.0	11
4.5	12
5.0	4

Review 4

1. Solve for x: $2 + \sqrt{3x + 7} = 15$ _____

2. Solve for x: $|7x + 11| = 16$ _____

3. Find the turning point and axis of symmetry: $-3x^2 + 7x = -11$

    tp: _____   axis of symmetry: _____

4. Describe the roots of the equation $-2x^2 - 3x + 7 = 0$.

---

5. Graph and find the solution set:
$x^2 - 5x + 11 < 7$

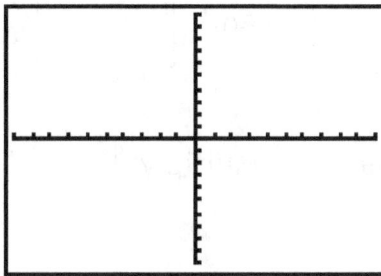

6. Graph and find the solution set:
$|-3x + 7| - 3 \geq 5$

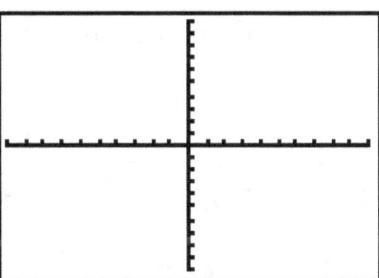

7. In 2003 Robert Sorlie won the Iditerod Trail Sled Dog Race with a time of 9 days, 15 hours, 47 minutes, and 36 seconds. In 2002 Martin Busen won with a time of 8 days, 22 hours, 46 minutes, 2 seconds. How much faster was Busen than Sorlie?

    _____

8. Write without an exponent: i^{78} _____

9. Write without an exponent: i^{345} _____

10.   Add: $3 - 4i$ and $-5 + 3i$ _____

11.   Subtract: $1 + 5i$ and $-2 - 7i$ _____

12.   Multiply: $8-i$ and $-3+2i$ _____

13.   Multiply: $3-4i$ and $3+4i$ _____

14.   Find the complex conjugate: $-12+4i$ _____

15.   Find the complex conjugate: $4-10i$ _____

16.   Evaluate: $|2-5i|$ _____

17.   Evaluate: $|-6+4i|$ _____

18.   Add: $\begin{bmatrix} -2 & 3 \\ 1 & 7 \end{bmatrix} + \begin{bmatrix} 3 & -5 \\ 4 & -3 \end{bmatrix}$ _____

19.   Multiply: $\begin{bmatrix} -2 & 0 & 5 \\ -1 & 3 & 7 \end{bmatrix} * \begin{bmatrix} 4 & -1 \\ -3 & 2 \\ 7 & 5 \end{bmatrix}$ _____

20.   Find the inverse: $\begin{bmatrix} -2 & 3 & 5 \\ 4 & -5 & 1 \\ 0 & -3 & 7 \end{bmatrix}$ _____

21.   Does $\begin{bmatrix} 2 & -5 \\ 6 & 9 \\ -2 & 4 \end{bmatrix}$ have an inverse?   _____

22.   What types of matrices have inverses?

----

23.   What types of matrices can be added?

----

24.   What types of matrices can be multiplied?

----

25.   Solve: $\begin{array}{l} 4x - 11y = 15 \\ -3x + 4y = 21 \end{array}$

_____

26.   Solve: $\begin{array}{l} \dfrac{1}{2}x - \dfrac{2}{3}y = \dfrac{5}{12} \\ \dfrac{-1}{4}x + 3y = \dfrac{7}{8} \end{array}$

_____

27.   A toy rocket is following the path $y = -16x^2 + 150x$, where y is the rocket's height in feet and x is the time in seconds. Show the path the rocket takes on the grid below.

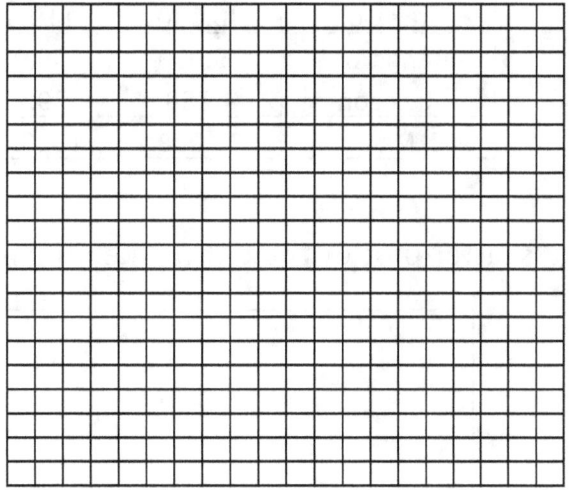

   a. After how many seconds, to the nearest hundredth, will the rocket land? _____

   b. What is the maximum height the rocket reaches?

   _____

28.   A school is purchasing graphing calculators. One teacher has requested 10 TI-83+'s and 5 TI-83+ Silvers for a total of $1460.75. Another teacher has requested 15 TI-83+'s and 3 TI-83+ Silvers for a total of $1686.00. What is the price of one TI-83+? _____ What is the price of one TI-83+ Silver? _____

29.   The table below gives the body mass in kilograms and the brain mass of selected mammals.

Mammal	Body Mass in kg	Brain Mass in grams
African Elephant	6654.000	5712.0
Arctic Fox	3.385	44.5
Baboon	10.550	179.5
Cat	3.300	25.6
Cow	465.000	423.0
Giraffe	529.000	680.0
Goat	27.660	115.0
Golden hamster	0.120	1.0
Gray seal	85.000	325.0
Gray wolf	36.330	119.5
Kangaroo	35.000	56.0
Raccoon	4.288	39.2
Man	62.000	1320.0

a. Create a scatter plot of the data. Sketch and label below.

b. Calculate the line of best fit for the data: _____

c. What is the linear correlation coefficient? _____

d. Calculate the power regression for the data: _____

e. Is this a better fit than the linear regression? _____ Explain:

_____

_____

30.   The table below gives the statistics for the American League pitchers with the lowest earned run averages after one month of play.

Pitcher	Walks	Strike Outs	Earned Run Average
Groom	2	7	0.00
B Ryan	4	13	0.00
D Baez	4	4	0.00
R Lopez	3	8	0.00
Malaska	3	3	0.00
Adkins	3	4	0.00
Stanford	5	5	0.82
Fultz	2	6	1.08
Gaudin	3	5	1.17
Westbrook	4	8	1.46
Sabathia	11	18	1.71
Adams	7	9	1.86
Mecir	1	8	1.93
Levine	2	5	1.93
Duchscherer	3	2	1.93
Ramirez	0	2	1.93
J Roa	1	6	2.08
T Hudson	4	16	2.15

a. Find the following for the earned run averages:

   i.   Mean   _____

   ii. Lower Quartile   _____

   iii. Median   _____

   iv. Upper Quartile   _____

   v.   Standard Deviation _____

b. Find the variance in the number of walks. _____

c. Find the mean absolute deviation in the number of strikeouts.

                                              _____

31. Barry Bonds is hitting .525.
    a. What is the probability that he will go 4 for 4 in his next game?

    _____
    b. What is the probability that he will have at least 3 hits in 5 at bats?
    _____
    c. What is the probability that he will have at most 2 hits in 5 at bats?
    _____

32. What is the 4th term in the expansion of $(x + y)^6$?

    _____

33. What is the 3rd term in the expansion of $(a + 4)^5$?

    _____

34. A coin is weighted so that the probability of landing on heads is .8. What is the probability that the coin will land on tails 3 out of 4 tosses?

    _____
35. After one month of play, the American League batters had a mean of 55.8 at bats with a standard deviation of 9.6. Assuming a normal distribution,
    a. What percent of the batters have between 50 and 60 at bats?
    _____
    b. What percent of the batters have between 40 and 52 at bats?
    _____
    c. How many at bats would a batter at the 75th percentile have?

    _____

Review 5

1. Convert to radians: $75°$ _____

2. Convert to radians: $330°$ _____

3. Convert to degrees: $\dfrac{3\pi}{2}$ _____

4. Convert to degrees: $\dfrac{\pi}{7}$ _____

5. A spider creates a web connecting the opposite vertices of a stop sign. What is the measure, in degrees, between the sections of the web? _____ What is the measure, in radians, between the sections of the web?

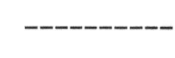

     _____

6. A 9-inch pie and an 8-inch pie are each divided into 6 equal pieces. What is the ratio of the measure of the outer crust of a piece of the smaller pie to the measure of the outer crust of a piece of the larger pie?

     _____

7. Graph $y = 2x^2 - 3x - 7$ in parametric mode and sketch below.

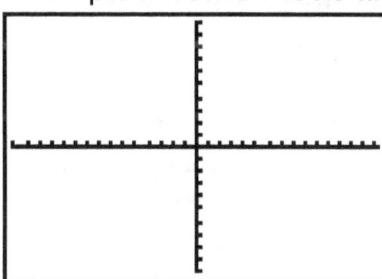

8. Graph $y = -4x^2 + 2x + 3$ in parametric mode and sketch below.

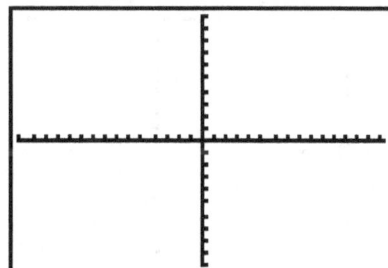

9. Graph $(x+3)^2 + (y-1)^2 = 9$ in parametric mode and sketch below.

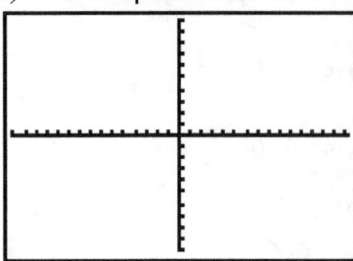

10. Graph $(x-2)^2 + (y-5)^2 = 25$ in parametric mode and sketch below.

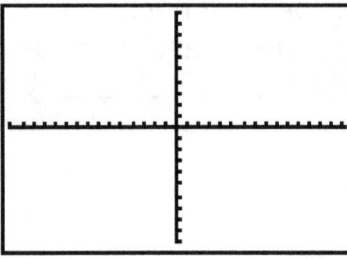

11. Graph $\dfrac{x^2}{4} + \dfrac{y^2}{49} = 1$ in parametric mode and sketch below.

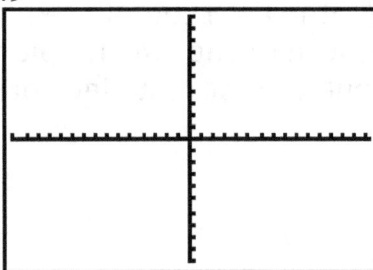

12. Graph $9x^2 + 16y^2 = 144$ in parametric mode and sketch below.

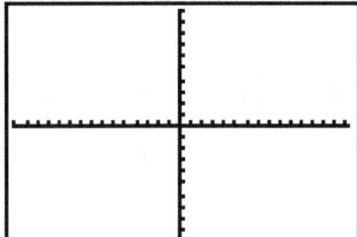

13. Graph  $\dfrac{x^2}{9} - \dfrac{y^2}{100} = 1$ in parametric mode and sketch below.

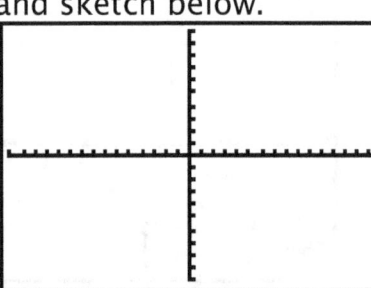

14.   Graph $4x^2 - 25y^2 = 100$ in parametric mode and sketch below.

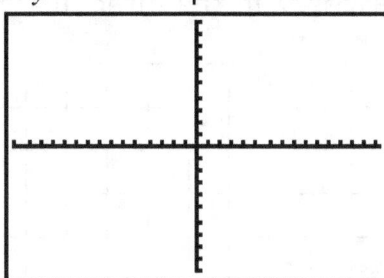

15.   What is the amplitude of the sine curve below? (y-scale=1)

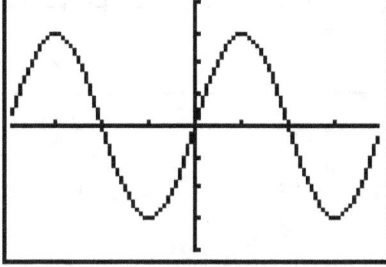

    _____

16.   What is the amplitude of the cosine curve below?(y-scale=1)

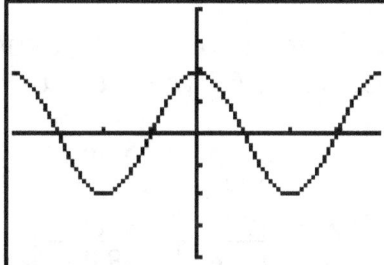

    _____

17.   What is the frequency of the equation $y = 4\sin\dfrac{1}{3}x$? _____

18.   What is the frequency of the equation $y = 2\cos5x$? _____

19.   Graph $y = 3\sin2x$ below. Use the interval $0° \le x < 360°$

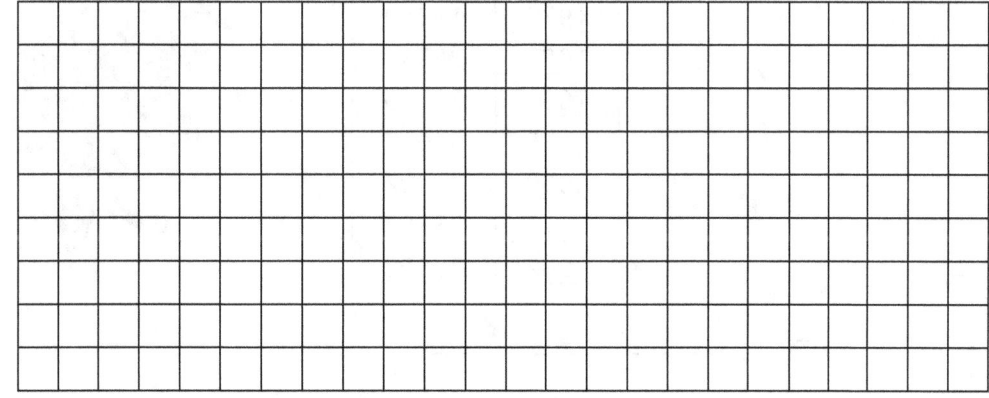

20. Graph $y = \cos\dfrac{1}{2}\theta$ in the interval $-\pi \le \theta \le \pi$.

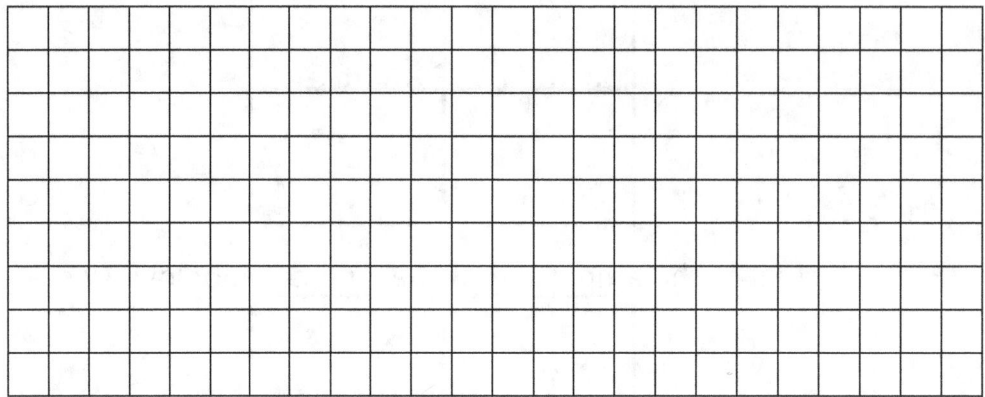

21. How will the graph of $y = \sin x - 2$ differ from the graph of $y = \sin x + 3$?

_____

_____

22. Use the data below to find an exponential model for the number of US residents 65 years old or older.

Year	Population ≥ 65 (in thousands)
1900	3080
1910	3949
1920	4933
1930	6634
1940	9019
1950	12269
1960	16560
1970	19980
1980	25550
1990	31079
1995	33619
2000	34992
2001	35353
2002	35602

$y=$ _____

23.　　　Use your model in #22 to estimate the population greater than 65 years old in

    a. 2010 _____

    b. 2020 _____

    c. 2050 _____

24.　　　Find $\log_3 5$　　　　　　　　　　　_____

25.　　　Find $\log_6 18$　　　　　　　　　　_____

26.　　　Solve for x: $\log_2 x = 4$　　　　　_____

27.　　　Solve for x: $\log_5 x = 3$　　　　　_____

28.　　　Solve for x: $\log_x \dfrac{1}{8} = -3$　　　　_____

29.　　　Graph $y = \log_3 x$ and sketch below.

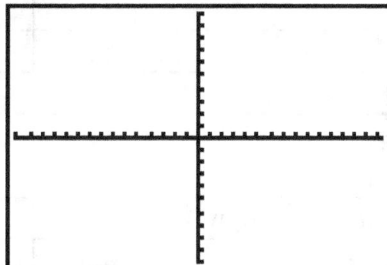

30.　　　If $f(x) = 3x^2 - 2$ and $g(x) = 4x + 1$, find

    a. $(f \circ g)(-2)$　　　　　　　　　_____

    b. $(g \circ f)(-2)$　　　　　　　　　_____

    c. The rule describing $(f \circ g)(x)$

                    _____

31.　　　If $g(x) = \dfrac{x}{2} - 5$, find $g^{-1}(x)$　　_____

32.　　　If $h(x) = 3x + \dfrac{1}{4}$, find $h^{-1}(x)$　　_____

33.　　　If Kate borrows $5,000 for college at the beginning of her first year, how much will she have to pay back at the end of her fourth year if the interest is 4.6% compounded annually?

                    _____

34.　　　If Rob buys a truck for $15,550, what will the value of the truck be in 3 years?

                    _____

The *Watertown Daily Times* published the following table on May 13, 2004:

Gasoline Prices - Watertown

	Lowest Price Per Gallon	
Station	May 7, 2004	May 12,2004
Agway Energy Products	1.92	2.04
Byrne Dairy	1.93	2.04
Fastrac Markets	1.98	2.06
General Store	1.94	2.04
Griff's Mini Mart	1.97	2.08
Hess Service Station	1.99	2.08
Interstate Mobil Mart	2.03	2.07
Just N Case	1.99	2.06
Mercer's	1.99	2.08
Nice-N-Easy	1.99	2.06
Russell's	2.00	2.11
Ryan's Jet Gas	1.94	2.04
Stewart's Shops	1.97	2.06
Sugar Creek Stores (Mobil)	1.96	2.06
24-Hour Mini Mart	1.98	2.06
Sunoco	1.98	2.06

35. Find

    a. The Mean for each date   _____ _____

    b. The Median for each date   _____ _____

    c. The Standard Deviation   _____ _____

    d. The Mean Absolute Deviation _____ _____

    e. The Range   _____ _____

    f. The Variance   _____ _____

36. Find the linear relationship between the May 7th prices and the May 12th prices.   _____

    How strong is this linear relationship? Explain using the correlation coefficient.

_____

_____

37. The Punxsutawney Groundhog Club has a record of
    Punxsutawney Phil's activities for each
    Groundhog's Day since 1887. His tally to date is:
           Saw shadow: 93
           No shadow: 14
           No record: 9

   (Do not include years for which there was no record.)
What is the probability that Phil will see his shadow in 2005?

What is the probability that Phil will not see his shadow at least 3
of the next 5 years?

                                             ----

38. In 2003, the New York Yankees had a winning percentage of
.623.
   What is the probability that they won all three games of a
randomly chosen 3-game series?

                                             ----

39. Find the inverse of $\begin{bmatrix} -2 & 3 & 11 \\ 8 & -4 & 5 \\ 0 & -1 & 6 \end{bmatrix}$. Convert any decimals to

fractions.

                                          ----

40. If 4 geraniums and 3 rose bushes cost $33.37 before tax, and 5
    geraniums and 7 rose bushes cost $61.18 before sales tax, what
    will   Karen pay for 2 geraniums and 4 rose bushes including a
    sales tax of 7¼%?

Final Exam Reveiw

1. The roots of $3x^2 - 5x - 2 = 0$ are
    a. Real, rational, equal
    b. Real, rational, unequal
    c. Real, irrational, unequal
    d. Imaginary

    _____

2. Use the quadratic formula to find the roots of $10x^2 + 3x - 7 = 0$.

    _____

3. Solve for x: $\sqrt{3x+7} = x + 2$ _____

4. Find the axis of symmetry and turning point of the equation
    $y = -2x^2 + 3x - 4$.   _____   _____

5. On Late Night With David Letterman, Dave sometimes has objects thrown from a five-story building just to see what happens when they hit the pavement below. A five-story building is approximately 55 feet tall. Find an equation that models the path of a melon thrown from the top of the building if it reaches a maximum height of 57 feet one foot away from the building and lands 12 feet from the building.

    $y=$ _____
    Graph your model on the grid below.

6. Graph $y = |3x - 5| + 4$

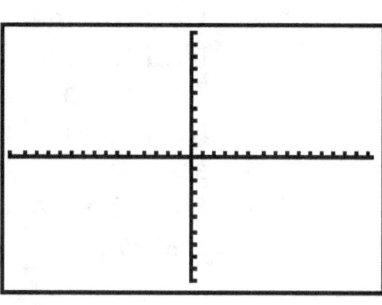

7. Change the equation in #6 so that the turning point lies on the *x*-axis.

8. Solve by graphing: $-x^2 + 3x + 7 \leq 4$

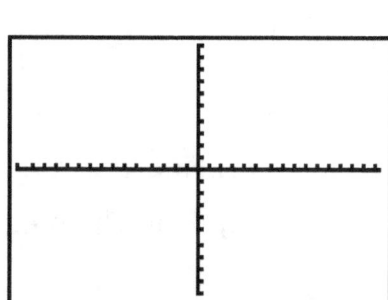

9. Solve by graphing: $|4x + 8| \geq 4$

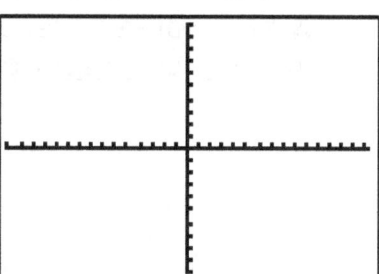

10. Use the data in the table below to answer the questions on the next page.

State	Tax on Gasoline (¢/gal)
Connecticut	25
Delaware	23
Maine	24.6
Maryland	23.5
Massachusetts	23.5
New Hampshire	20.6
New Jersey	14.5
New York	22.6
Pennsylvania	25.9
Rhode Island	31
Vermont	20
Virginia	18.1

a. Find the mean tax per gallon: _____
b. Find the median tax per gallon: _____
c. This set of data contains only the tax for states in the Northeast. Find the standard deviation based on all fifty states: _____
d. Find the standard deviation for just the Northeastern states:

_____

e. Create a box-and-whisker plot for the data given.

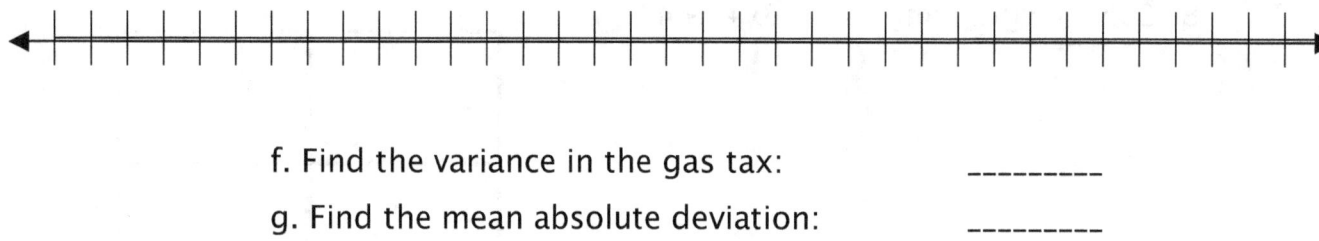

f. Find the variance in the gas tax: _____

g. Find the mean absolute deviation: _____

11.   The table below gives the price per gallon for gasoline in the Syracuse area as reported by customers to the website www.syracusegasprices.com as of June 4, 2004.

Price Per Gallon	Frequency
$2.05	1
2.07	1
2.08	2
2.09	5
2.11	6
2.15	7
2.16	3
2.17	3
2.19	1
2.23	1

a. Find the mean: _____

b. Find the median: _____

c. Find the variance: _____

d. Find the mean absolute deviation: _____

12. The New York State Department of Environmental
     Conservation monitors precipitation and acid rain at
     locations around the state. At Wanakena, from 1987 to
     2001 the mean precipitation was 42.31 inches with a
     standard deviation of 4.59 inches. The pH level of the
     precipitation had a mean of 4.39 with a standard
     deviation of .07. The precipitation and pH levels
     approximate a normal distribution.

     a. Find the probability that next year the precipitation will be
         at least 48 inches.   _____

     b. Find the probability that next year the precipitation will be
         between 40 and 45 inches.   _____

     c. Find the probability that the pH level will be below 4.3.

                 _____

     d. Find the probability that the pH level will be above 4.6.

                 _____

     e. Find the amount of precipitation at the 33rd percentile.

                 _____

     f. Find the amount of precipitation at the 90th percentile.

                 _____

     g. Find the pH level at the 40th percentile.   _____

     h. Find the pH level at the 95th percentile.   _____

13. Unleaded gasoline prices in the Syracuse area have followed the
     trend displayed in the table below: (Note: "Today" is June 4, 2004)

	# of days ago	Price per Gallon
Today	0.1	$2.15
Yesterday	1	$2.13
One Week Ago	7	$2.13
One Month Ago	31	$1.87
One Year Ago	365	$1.53

Find each type of regression equation and correlation coefficient
relating the number of days ago to the price per gallon:

a. Linear   y= _____   r= _____

b. Logarithmic   y= _____   r= _____

c. Exponential   y= _____   r= _____

d. Power   y= _____   r= _____

e. Why was the number of days ago for today entered as 0.1?

_____

_____

f. Which is the best model? _____ Why?

_____

_____

g. According to the "best" model, what will be the price of gas 30 days from "Today" ?   _____

h. What will be the price of gas one year from "Today"?   _____

i. What will be the price on Labor Day of 2004? (Use Days Between Dates to find the number of days between June 4 and Labor Day.)

_____

14. Find an exponential model for the federal debt, based on the data in the table. Let x=0 correspond to 1960.

Accumulated Gross Federal Debt

Year	Amount (in billions of $)
1960	291
1965	322
1970	381
1975	542
1980	909
1985	1818
1990	3207
1995	4921
1996	5182

y= _____

Use the model to predict the federal debt in 2005.

_____

15. President Ronald Reagan died on Saturday, June 5,2004. President Richard Nixon died on April 22, 1994. How much longer, to the nearest day, did Ronald Reagan live?   _____

16. The first women's winner of the Ironman Triathon was Lyn Lemaire of the US with a time of 12 hours, 55 minutes, 0 seconds. In 2002, Natascha Badman of Switzerland was the women's winner with a time of 9 hours, 7 minutes, and 54 seconds. How much faster was Natascha than Lyn?

        ----

17. Solve using a matrix:
$$4x+6y= -34$$
$$2x+3y= -17$$

        ----

18. Two families go for soft ice cream after a softball game. If the Nortz's spend $10.25 for 3 medium cones and 2 sundaes and the Lee's spend $8.75 for 4 medium cones and one sundae, what would another customer have to pay for 2 medium cones and 2 sundaes?

        ----

19. Find the inverse of $\begin{bmatrix} -4 & 1 & -2 \\ 0 & 3 & 6 \\ -1 & 0 & 7 \end{bmatrix}$.

        ----

20. If the New York Yankees currently have a winning percentage of .636, what is the probability that they will win at least 4 of their next 5 games?

        ----

21. Find the 3rd term in the expansion of $(x-2y)^5$

        ----

22. Find $\log_3 8$.

        ----

23. Find x: $\log_5 x = 25$

        ----

24. Find the product of $3-2i$ and $6+7i$.

        ----

25. Find $|4+12i|$.

        ----

26. Find the conjugate of $-12+5i$.

        ----

27. Simplify i^{23}.

        ----

28. Simplify i^{40}.

        ----

29.   Change to radians:
     a. 360°   _____

     b. 45°   _____

     c. 125°   _____

     d. 210°   _____

30.   Change to degrees:
     a. 4π _____

     b. 3π _____

     c. $\pi/2$ _____

     d. $2\pi/3$ _____

     e. π _____

31.   If you deposit $5000 in a savings account that pays 3% interest, compounded annually,
     a. How much money is in the account after nine years?

                                  _____

     b. When will the balance reach $50,000?

                                  _____

32.   The world population in 1950 was about 2.5 billion people and has been increasing at approximately 1.85% per year.
     a. Find the world population this year.

                                  _____

     b. In what year will the population be double this year's?

33.   If Ray buys a computer this year for $1750, what will the depreciated value of the computer be 4 years from now?

                                  _____

34.   How long will it take to double an investment of $100 at 6.5% interest if the interest rate is compounded continuously?

                                  _____

35.   If interest is compounded continuously, what annual rate must you receive if your investment of $1500 is to grow to $2100 in six years?

                                  _____

36. If $f(x) = 3x - 10$ and $g(x) = -x^2 + 4$, find
    a. $f(3)$ _____
    b. $g(-2)$ _____
    c. $g(f(x))$ in simplest form _____

37. If $h(x) = \dfrac{2}{3}x - 9$, find $h^{-1}(x)$ _____

38. Graph $y = \sin x$ from $x = -\pi$ to $x = \pi$.

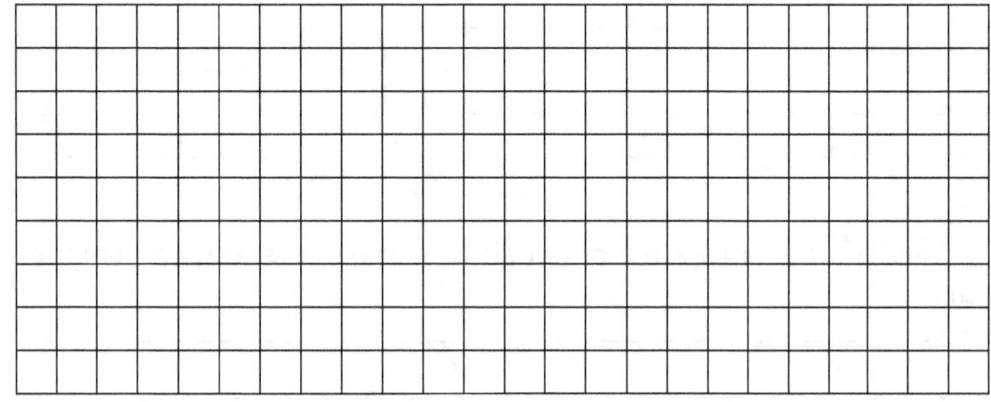

39. Graph $y = \sin x + 3$ $x = -\pi$ to $x = \pi$.

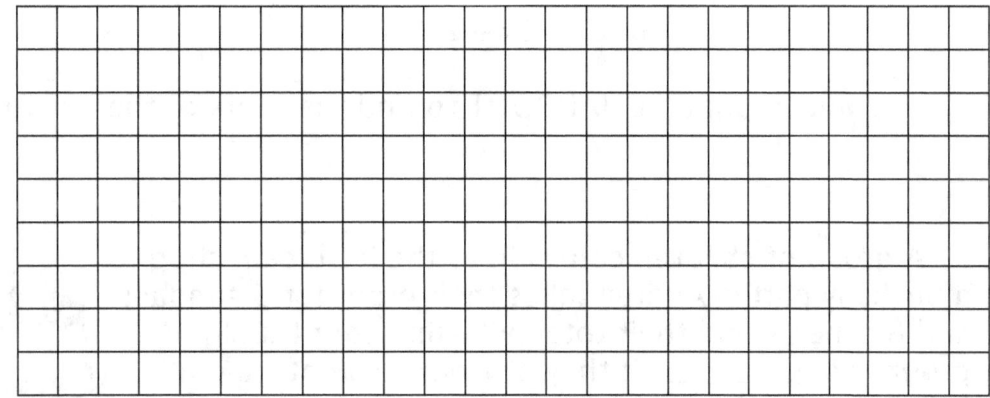

40. Explain what adding 3 did to the graph of $y = \sin x$.

_____

_____

41. What is the period of the function $y = -3\sin 5x$?
    a. In degrees. _____
    b. In radians. _____

42. Graph $y = \cos x$ in the interval $0° \leq x \leq 360$.

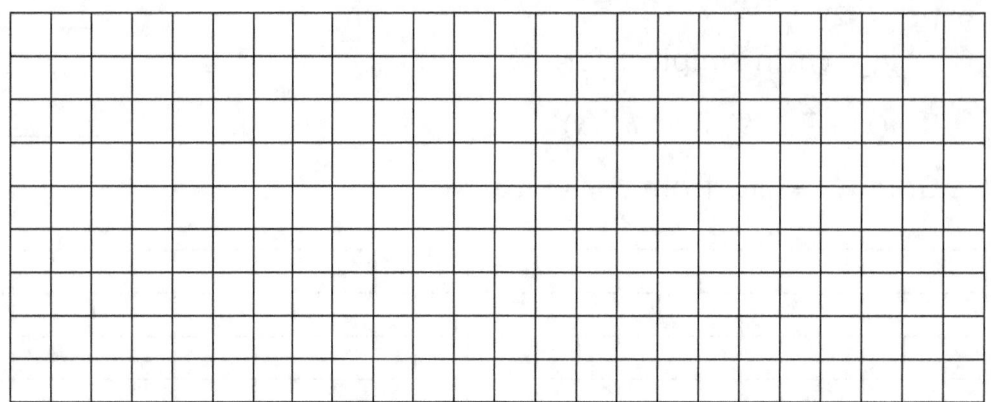

43. Explain the difference between the sine curve and the cosine curve.

   ----

44. Use your program QUADROOT to find the roots of the equation

   $0 = 2x^2 - 2x + 2$

   _____   _____

45. Use your program QUADROOT to find the roots of the equation

   $0 = x^2 - x + 15$

   _____   _____

46. A group of children comes in to the ice cream shop for a birthday party. Andrea takes their order for 5 sundaes and 8 cones. Find their total bill using your ICECREAM program? _____ If they pay with a twenty and a five and tell her to keep the change, what amount have they tipped her? _____ What percent was the tip of the total bill? _____

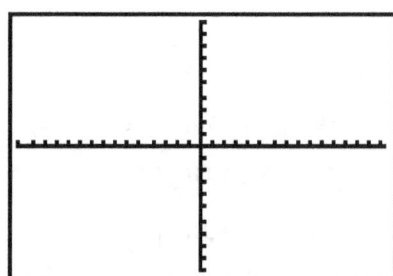

47. Graph in parametric mode and sketch: $16x^2 + 36y^2 = 576$

48. Graph in parametric mode and
    sketch: $x = 2y^2$

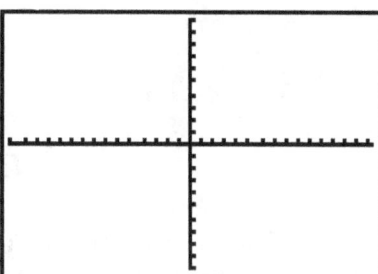

49. Sketch the figure BEAR with vertices
    B(4,1), E(4,7), A(6,4), R(2,3).

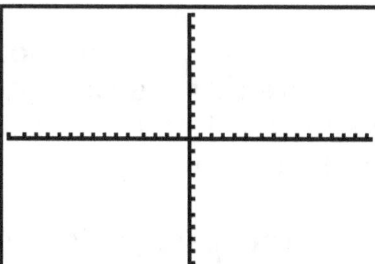

50. Sketch the figure B'E'A'R', the image
    of BEAR from #49 after a reflection in
    the line $y = -x$.

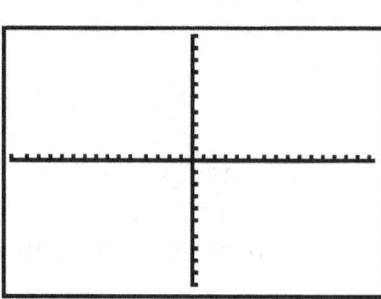

51. Evaluate: $\displaystyle\sum_{x=-1}^{4}(4x^2 - 2x)$ _____

52. Evaluate $\displaystyle\frac{1}{5}\sum_{n=1}^{4}(3n-7)^{(2-n)}$ _____

Appendix

Using TI-Graph Link

The TI-Graph Link allows communication between your TI graphing calculator and your PC.

It is necessary for downloading applications from the TI website, but be aware that the computer downloads the application to itself first then sends it to the calculator. If your school has placed restrictions on the computer you may not be able to download the applications.

1. The first step is to load the software that comes with the graph link on your computer. This will require permission from your system operator if there are restrictions on your computer.

2. Connect the graph link to your hard drive.

3. Connect the graph link to your calculator as though you were linking to another calculator. Be sure the connection is complete. You can usually feel a "click" when the connection is made.

**Note: It is always best to plug your calculator in and click on the TI-Graph Link program before setting up your screen if you are printing the screen. The calculator will usually return to the home screen when the program is started.

4. To insert your calculator screen in a document:
   a. Set up the page you will be inserting the screen into.
   b. Set up the calculator to show the screen you wish to insert.
   c. With the traditional software you can choose Link and Get Screen or click on the picture of a camera.
   d. With the TI Connect software choose screen capture.
   e. With each type of software you are given the option of _**adding**_ a border. This will make the picture look more screen-like on your page. (Go to Image in TI Connect)
   f. Copy the screen to the clipboard.
   g. Return to your word-processing document. The picture is available from the clipboard.
   h. You will need to close the screen you have just made before transferring a new one.

** Note: Some areas of the calculator's program are easier to transfer than others. Depending on the amount of memory involved and the version of the software, it may take several minutes to copy a screen. Some screens are not possible to copy.

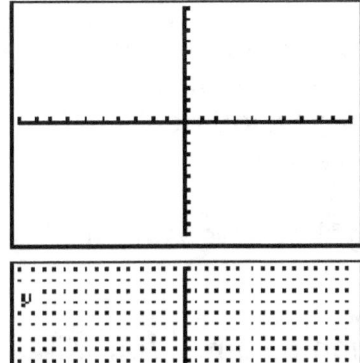

Standard coordinate axes can be inserted into a page to create a place to sketch a graph by hand.

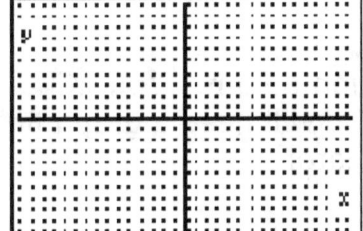

This screen was made with the grid and labels turned on in FORMAT.

Create graphs or charts for reports.

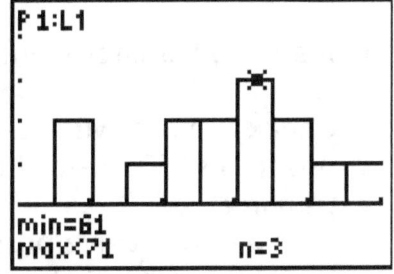

Show window dimensions and graphs for assignments.

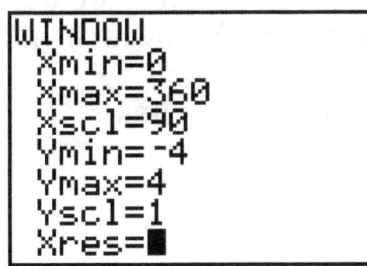

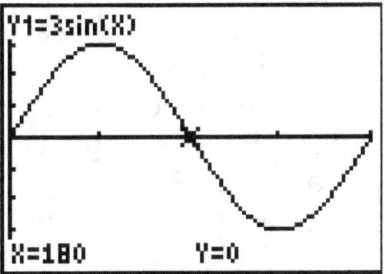

Updating the Operating System

The TI-83+/TI-84+ is actually a handheld computer. It can be "upgraded" just like your computer.

As Texas Instruments develops new accessories and applications for the TI-83+/TI-84+, improvements are made to the operating system to make the calculator compatible with the new developments. This saves you having to buy a new calculator every year or so.

The operating system can be downloaded from the TI website or transferred from another calculator.

To download from the TI website go to education.ti.com. You will save the operating system to your computer and will need a graph-link to download from your computer to your calculator.

It is faster to transfer from a calculator that already has the newer OS.

1. Check the receiving calculator to see if there are unarchived programs. These will be lost unless they are archived before the transfer.
    a. To archive programs, press 2nd, + and choose 5:Archive.
    b. "Archive" should now appear on the home screen. Press PRGM and choose the name of the program you want to archive.
    c. Press ENTER. The program will now be safe during the OS update. Applications are automatically archived and will be safe.
2. Link the two calculators with the unit-to-unit link.
3. Prepare the receiving calculator first:
    a. Press 2nd LINK (the "X,T,θ,n key).
    b. Highlight RECEIVE.
    c. Press ENTER.
4. On the sending calculator:
    a. Press 2nd LINK.
    b. Scroll down the SEND list to G:SendOS
    c. Press ENTER.
5. Respond to any prompts.
This will take several minutes to complete.

Formulas

Area of Triangle

$K = \frac{1}{2} ab \sin C$

Law of Cosines

$a^2 = b^2 + c^2 - 2bc \cos A$

Functions of the Sum of Two Angles

$\sin (A + B) = \sin A \cos B + \cos A \sin B$
$\cos (A + B) = \cos A \cos B - \sin A \sin B$

Functions of the Double Angle

$\sin 2A = 2 \sin A \cos A$
$\cos 2A = \cos^2 A - \sin^2 A$
$\cos 2A = 2 \cos^2 A - 1$
$\cos 2A = 1 - 2 \sin^2 A$

Functions of the Difference of Two Angles

$\sin (A - B) = \sin A \cos B - \cos A \sin B$
$\cos (A - B) = \cos A \cos B + \sin A \sin B$

Functions of the Half Angle

$\sin \frac{1}{2} A = \pm \sqrt{\dfrac{1 - \cos A}{2}}$

Law of Sines

$\dfrac{a}{\sin A} = \dfrac{b}{\sin B} = \dfrac{c}{\sin C}$

$\cos \frac{1}{2} A = \pm \sqrt{\dfrac{1 + \cos A}{2}}$

Normal Curve

Standard Deviation

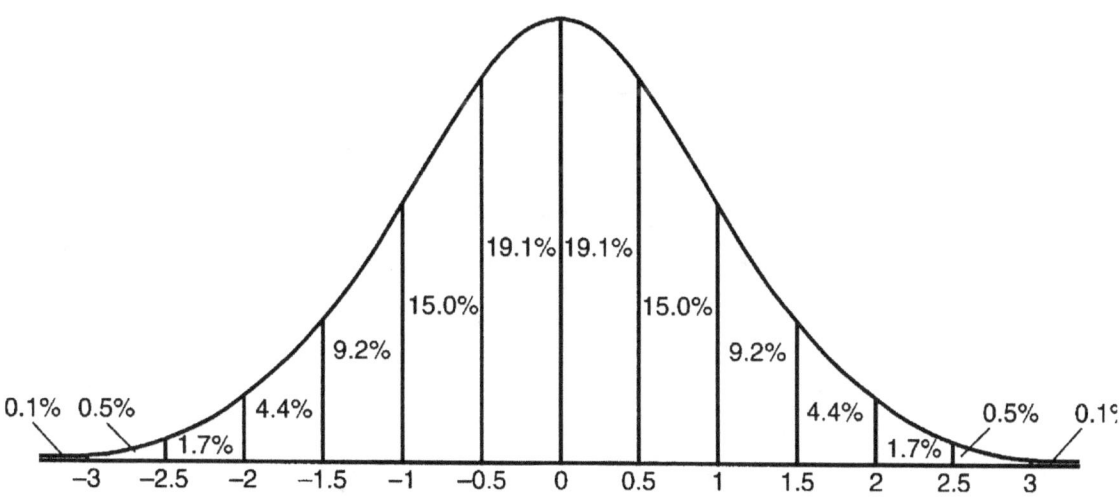

www.ingramcontent.com/pod-product-compliance
Lightning Source LLC
Chambersburg PA
CBHW081108170526
45165CB00008B/2366

* 9 7 8 1 4 2 5 1 0 2 1 6 6 *